Species, Serpents, Spirits, and Skulls

Species, Serpents, Spirits, and Skulls

Science at the Margins in the Victorian Age

Sherrie Lynne Lyons

Cover image: "The Great Serpent (According to Hans Egrede)." General Research Division. The New York Public Library. Astor, Lenox, and Tilden Foundations.

Published by State University of New York Press, Albany

Printed in the United States of America

For information, contact State University of New York Press, Albany, NY
www.sunypress.edu

Production by Eileen A. Meehan
Marketing by Anne M. Valentine

Library of Congress Cataloging-in-Publication Data

Lyons, Sherrie Lynne, 1947–
Species, serpents, spirits, and skulls : science at the margins in the victorian age / Sherrie Lynne Lyons.
p. cm.
Includes bibliographical references and index.
ISBN 978-1-4384-2797-3 (hardcover : alk. paper)
1. Science—Great Britain—History—19th century. 2. Great Britain—Intellectual life—19th century. I. Title.
Q127.G4L96 2009
509.41'09034—dc22

2008054150

10 9 8 7 6 5 4 3 2 1

For Cassandra and Grahame

Contents

Illustrations

Table

Figures

Preface

We live in a culture where science permeates virtually every aspect of our lives. Not only does it affect how we think about broad areas of research, from star wars to the war on cancer, it has also invaded the most private part of our everyday lives. Science has shaped our views on everything from what we should have for breakfast to how we should be raising our children to what our sex life should be like. Yet most people are woefully ignorant of what distinguishes science from other types of knowledge and they lack any sense of the historical development of scientific ideas. Often yesterday's heresy is today's science (and conversely today's pseudoscience is yesterday's science).

This study uses four historical cases—sea serpent investigations, phrenology, and spiritualism, along with Darwin's theory of evolution—to address issues of scientific marginality and legitimacy. This comparative approach illustrates that distinguishing between credible and doubtful conclusions in science is not as easy as the popular image of science tends to suggest. As David Knight has pointed out, science "is and always was based on a judicious mixture of empiricism and faith," and thus, in the midst of discovery it is often difficult to distinguish what constitutes science from what does not.[1] Therefore, the history of sciences that failed to become established has many lessons to teach us. Why do certain topics become subjects of scientific investigation at a particular moment in history? Why do some topics remain at the margins and others eventually become known as pseudoscience, while others grow and prosper, often spawning whole new disciplines? In fact, the distinction between science and pseudoscience is not sharp. The term "scientist" did not even exist until the middle of the nineteenth century. Rather than the term "pseudoscience," which literally means false science, a more appropriate term to describe my case studies is "marginal" science. Each one attracted the interest of prominent scientists as well as the general public. Nevertheless, they all remained at the edges of scientific respectability. In contrast, evolutionary theory, while initially regarded as scientific heresy, rather quickly became the new scientific orthodoxy. The differing fates of each of these subjects are the topic of this book.

Although these rather different episodes have aspects that are unique, certain common themes emerge. Each tells a complex story of professionalization as various disciplines emerged, each vying for power and authority. Social as well as cognitive factors played a significant role in the gathering and interpreting of evidence. "Facts" were used to support a variety of different positions, and ideas were often accepted with very little evidence, and held on to despite contradictory evidence. In spite of this, these episodes also illuminate how scientific practice has moved us to an ever increasingly accurate view of the material world. The final chapter explores why this is so and how the boundary disputes of the Victorian era have shaped the practice of science today. The production of scientific knowledge is embedded within a social, political, and historical matrix. However, rather than being a hindrance, that very intimate relationship between science and society contributes to the richness and diversity of scientific ideas. Examining two present-day controversies that are connected to the case studies illustrates that boundaries continue to be negotiated, and that disagreement and debate is essential to scientific progress.

Acknowledgments

This book has been a long time in coming and evolved into something quite different than the original manuscript. It had its genesis while I was working on my doctorate on Thomas Huxley and found a file on sea serpents in Charles Lyell's scientific papers. David Jablonski encouraged me to apply for a small grant from Sigma Xi to look further into Lyell's interest in sea serpents. The project would never have gotten off the ground if it hadn't been for Robert Brugger, who, after reading about my talk on sea serpents, encouraged me to develop an entire manuscript. The project soon expanded far beyond sea serpents, and it benefited enormously from suggestions by Martin Fichman, Hy Kuritz, Bernard Lightman, and Tim Madigan who read an earlier version of the entire manuscript. Thank you to them and also to several anonymous reviewers who made helpful suggestions.

There have been many friends and colleagues along the way who have been a source of stimulating discussions as well as encouraging words. These include Garland Allen, Toby Appel, Ellen Banks, David Blitz, Jim Bono, Geoffrey Cantor, Claudine Cohen, James Elwick, Maria Gogarten, Zena Goldenberg, Gwen Kay, Carolyn Kirdahy, Peter Kjaergaard, Ricki Lewis, Eric Luft, Dan MacShea, Adrienne Mayor, Robert Perlman, Robert Richards, Martin Rudwick, Nancy Slack, Michael Sokal, Sally Stebbins, Marc Swetlitz, and Torbin Wolff. Thank you to Myrna Becker for helping with the index and final corrections. A special thanks to John Greene who helped me in so many ways, but sadly is no longer with us. Several paleontologists—including Paul Sereno, Jack Horner, Norman McLeod, Mark Norell, and Tim Tokaryk—took the time to discuss with me the relationship of fossils to mythology. Most of all, I thank my good friend and colleague Douglas Allchin who read numerous versions and understood what I was trying to do: write a scholarly book but for a broader audience. This has been a daunting task. The final chapter in particular owes a great deal to Douglas, not only in providing relevant references but also in shaping the organization and tightening the philosophical arguments.

As always, the author is responsible for the final result and any mistakes are my own. I thank James Peltz at State University of New York Press for believing in this manuscript and Dee Quimby and her team at A&B Typesetters and Editorial Services for helping bring it to fruition.

It was a pleasure and tremendous help to work at the Center for Inquiry as a visiting scholar in residence. The library there specializes in materials that were especially relevant for this project. The librarians at all the institutions I worked at were always extremely helpful in assisting me in finding material. I thank all of them for their help at the Wellcome Library for the History of Medicine, the British Museum of Natural History, the British Museum at Colindale, Edinburgh University Library, and the library at the spiritualist community of Lily Dale, New York. Financial support was received from Sigma Xi and National Science Foundation Award #9906085.

Finally, I thank my wonderful children, Cassandra and Grahame, who continue to be a source of joy and pride.

1

Introduction

An Age of Transition

May you live in interesting times.

—Ancient Chinese curse

THE VICTORIAN ERA was indeed an interesting time, a time of stark contrasts. It was a time of hope and dismay, of great optimism coupled with anxiety, doubt, and fear. It has been called the "age of science" because of outstanding discoveries in a variety of different fields.[1] The theory of electromagnetism continued the grand unification of the physical sciences, building on the spectacular successes of the scientific revolution. Developments in natural history, geology, embryology, and taxonomy meant that in crafting his theory Charles Darwin could draw on a wealth of information that was unavailable to his predecessors. Victorians packed the lecture hall of the Royal Institute to hear lectures by Michael Faraday on electricity and magnetism and watched Humphrey Davy as he poured water into a model volcano filled with potassium. They listened attentively to John Tyndall's arguments against the possibility of spontaneous generation from inorganic matter. The enthusiasm they showed for Thomas Huxley's lectures on man's relationships to the lower animals caused him to quip to his wife, "By next Friday evening they will all be convinced that they are monkeys."[2] The age of science was also a time of unprecedented growth in the marginal sciences. Large numbers of Victorians experienced mesmeric trances, attended séances, and obtained phrenological evaluations of their character. Rather than the age of science, perhaps the Victorian period should be called the "age of contradictions," as countless numbers of Victorians became absolutely convinced of the reality of spiritual phenomena. However, these different interests are not contradictory. Rather, they reflect Victorians' hope that scientific advancements would make it possible to understand the human psyche, which in turn would allow them to come to terms with a rapidly changing society.

Victorian intellectuals themselves characterized the period as the "age of transition." For the first time in history, a population thought of their own time as an era of change *from* the past *to* the future.[3] Beginning in 1832, a series of reform bills were passed that would change the structure of British society, breaking down the rigid social hierarchy that doomed people to remain in the class they were born to. This breakdown of the old conception of status was not due primarily to ideas of democracy, but rather was economically driven. The development of commerce provided many new careers, allowing men to leave the land, dissolving the feudal hold wealthy landowners had on society.

Developments in science and technology fueled the industrial revolution, totally transforming the economic life of England. Instead of a strict system of fixed regulations determined by the rigid social hierarchy, we see the emergence of a laissez-faire capitalism. Many people believed that unbridled competition would weed out the less fit and lead to an overall improvement of society. Such ideas provided the backdrop to Charles Darwin's developing theory that in nature the constant struggle for existence resulted in selection of the most fit. As Herbert Spencer wrote, economic competition resulted in the "survival of the fittest" and Darwin adopted the phrase, realizing that a parallel process was also going on in nature.[4]

Such a dynamic society offered men the possibility of success both financially and socially that never before had existed for members of the middle and lower classes. However, progress did not come without a price. Throughout the Victorian period economic cycles of booms and busts occurred, resulting in large numbers of working-class people periodically meeting and agitating. Expanding business, developments in science, and the growth of democracy were sources of distress as well as satisfaction. From our perspective today, growth in democracy certainly seems like an unbridled good, but for many Victorians, fear of revolution went hand in hand with the idea of democracy. Across the channel, the French were recovering from the excesses of the French Revolution. The spread of both political and radical religious propaganda among the working classes suggested that Britain was not immune to the possibility of revolution.

The Victorian era was an exciting time filled with opportunities—opportunities that were not confined to political and economic spheres. Issues that had supposedly been settled for centuries were now open to vigorous debate. Religious beliefs, ethical theory, and human nature were all topics subjected to critical discussion and scrutiny. Was there a God or not? If so, was he a person or an impersonal force, or an indifferent force? Was there a heaven or hell, both or neither? If there was a true religion, was it theism or Christianity, and within Christianity—Catholicism or Protestantism? Did we have freewill or were we human automatons, and if we had the power of moral choice, what was its basis? Was it a God-given voice of conscience or was it the product of a rational calculating mind that decides what course of action will provide the greatest happiness for the greatest number of people? Was a human being just a more intelligent version of an ape? Public lectures, numerous societies, inexpensive editions of books, periodicals, newspapers, and pamphlets all provided forums for debate and meant that such discussions were not just confined to the intellectual elite. In such an environment of competing theories and beliefs not only were specific doubts raised about particular issues, but also the habit of doubt itself was unconsciously bred. This climate of uncertainty created a vague uneasy feeling, as individuals no longer felt secure in their own beliefs, and this in turn fueled the growth of the marginal sciences. Thus, the marginal sciences provide an important window into Victorian society. More significant for my purposes, they also demonstrate that the boundary between science, marginal science, and nonscience continually shifts as a variety of interrelated but distinct factors come into play. Each of the case studies examines the intersection between a specific body of developing scientific knowledge and the larger society revealing a complex intimate relationship between the two.[5]

In chapter 2, the story of the sea serpent shows that the line between fact and fantasy is not as clear as many scientists would have us believe. The discovery of fossil plesiosaurs and ichthyosaurs not only suggested a scientific basis for the sea monsters of ancient myth and legend, but also made plausible the possibility that the ancient creatures had survived to the present. Debates over the existence of sea serpents could be found in popular magazines as well as respected scientific journals. However, the sea serpent became a locus for boundary disputes as geology and paleontology struggled for scientific legitimacy. If the sea serpent were to be taken seriously, then its existence would have to be compatible with the prevailing views of earth history. But these views were in flux as a result of recent fossil finds and the serpent's existence or nonexistence was used to promote a particular point of view of earth history.

Yet sea serpent investigations remained marginalized for numerous reasons. One way that the emerging class of professional paleontologists and geologists could gain scientific authority was by making distinctions between amateurs and professionals. Amateurs made most of the serpent sightings, but the distinction between who was a professional and who was an amateur concerning matters of natural history was in part what was being contested. Another important aspect of the serpent story was defining what counted as evidence. The serpent literature reflected the language of the courts, drawing on eyewitness accounts and affidavits. In codifying what was regarded as scientific proof, sightings must eventually be backed up by actual specimens, which were never found. The collecting, marketing, and exhibiting of fossils played a critical role in both the popularization and professionalization of geology.[6] Furthermore, the existence of a relic surviving from the distant past became more and more untenable as a consensus view emerged concerning the history of life. The story of the sea serpent also illustrates that science has been as successful as it has by choosing carefully what questions it will investigate. Geologists and paleontologists decided it was not worthwhile to use their time and limited financial resources to search for a creature that might not even exist when a wealth of existing fossil and stratigraphical data awaited interpretation.

Chapter 3 examines the rise and fall of phrenology, continuing an exploration of many of the same issues that were raised in chapter 2. However, the history of phrenology is far more complex. Rather than revolving around essentially one scientific fact or theory: the existence of sea serpents, phrenology entailed an entire body of knowledge concerning the nature of mind and brain and the application of that knowledge. In addition, class, religion, and the threat of materialism played a significant role in how phrenology was received.

Franz Gall argued that the brain was the organ of the mind and should be studied accordingly. The investigation of mind had been the province of philosophers who claimed that the only way to understand the mind was through introspection. Gall challenged that assumption, pioneering techniques in neuroanatomy that have been invaluable. His naturalistic approach contributed to the separation of psychology from philosophy and paved the way for an evolutionary approach to the study of mind and behavior.[7] The phrenological connection to evolutionary theory again illustrates the blurring of boundaries between science and marginal science and the problematic nature of collecting and interpreting evidence.

Today, the importance of Darwin's ideas to biology and psychology remains central, but phrenology is regarded as pseudoscience. Gall at best receives sparse acknowledgment in histories of neurology and psychology. Yet Darwin's theory was relatively quickly accepted in part because Victorians had been exposed much earlier to the general idea of evolutionary change, and specifically to many ideas about the evolution of mind and behavior with the anonymous publication of *Vestiges of the Natural*

History of Creation in 1844.[8] Passages in Darwin's *The Descent of Man and Selection in Relation to Sex* show a striking similarity to those in *Vestiges*, in which the author Robert Chambers explicitly acknowledged his debt to Gall and phrenological doctrine.[9] Darwin used virtually the same kinds of evidence as Gall to show that humans were not an exception to his theory. Yet phrenology had a very different fate than evolution, as both of them become inextricably intertwined with the philosophical, religious, political, social, and scientific debates of the time.

One of the most striking aspects of the intellectual history of the nineteenth century was a faith that science could solve virtually all problems. Such an idea had been suggested in the Enlightenment, but never had much of a following. Victorians had great hope that the scientific assumptions and methodology that had been so spectacularly successful in the physical sciences could now be extended to the biological world including the lives of humans. The actions of human beings, like all other natural events, were subject to invariable laws that could be discovered. Advances in medicine would bring an end to physical suffering. Developments in technology promised to bring an end to poverty. The new science of sociology might even eliminate moral evil. Phrenology promised to be a truly scientific explanation of the mind, and with that offered the possibility of improving oneself, and therefore society.

Young physicians quickly recognized the potential of phrenology for increasing their prestige and authority and were the leaders of the phrenological movement in its early days. However, the old guard of the medical establishment along with the philosophers were not about to give up their authority easily and did everything they could to discredit it. But it would be wrong to conclude that phrenology never succeeded in becoming part of the scientific mainstream because of political maneuvering on the part of the university and medical establishment. Although Gall's basic premises were praised, the evidence for his theory of cerebral localization was torn to shreds. The advocates of phrenology were not interested in doing detailed anatomical work to further Gall's basic ideas and by the second half of the nineteenth century phrenology had lost virtually all of its reputable support.

However, the lure of phrenology was simply too strong and it took on a life of its own as countless societies sprung up. Its advocates believed that phrenology was the key to solving virtually all of society's ills and it allied itself with the various reform movements of the time. Phrenology at different times in its history was a mixture of theory and practice that represented a marginal science well on its way to becoming part of the scientific mainstream, but also degenerated into practices that were nothing short of quackery. At the same time it became a popular and powerful social movement that had many positive aspects to it and in that regard should be considered nonscience rather than marginal or pseudoscience.

Chapters 4 and 5 examine the many different factors that contributed to the widespread interest in spiritualism in all classes of society. Spiritualism in many ways epitomizes the often-conflicting tendencies that characterized Victorian society, particularly in regard to their attitude toward science and religion. Spiritualism spoke to the growing crisis in faith that permeated all classes of society. Victorians found their faith being challenged on a variety of fronts and the decline of Christianity with the threat of atheism caused anxiety on a variety of levels. Many people thought a loss of faith would result in a collapse of morality. By giving authority to the commandments and creating a fear of doing wrong, organized religion provided the sanction of moral obligation. Religious belief was necessary for both moral and social purposes. Therefore, if large numbers of people lost their faith, society would disintegrate. While challenges to religion were being raised at all levels of society, people were particularly worried about the spread of unbelief in the lower classes and was explicitly linked to a fear of revolution. This perceived connection between religious belief and respect for the law was clearly illustrated by various reviews of Darwin's *The Descent of Man* appearing in the most important newspapers. In arguing for a totally naturalistic account of humankind's origins, Darwin was condemned for "revealing his zoological anti-Christian conclusions to the general public at a moment when the sky of Paris was red with the incendiary flames of the Commune."[10]

Atheism was more common among members of the working class than any other group of people for a variety of reasons. First, Thomas Paine and other early freethinkers explicitly expressed religious skepticism in their radical writings. Second, the Church of England generally identified with and supported Tory and aristocratic principles and a conservative reading of the Bible that contributed further to prejudice against the church from the working classes. Finally, the suffering that the working classes endured seemed incompatible with the existence of a just and merciful God. One only had to turn to the serial installments of Charles Dickens's *Oliver Twist* in *Bentley's Miscellany* to realize that for many members of society life consisted of squalor and grinding poverty. The prevailing social order was under siege.

In addition to political and economic changes that led to a decline in the faithful, intellectual challenges from both within theology and outside of it made it increasingly problematic to accept a literal interpretation of the Bible. Although the Copernican revolution resulted in humans realizing that the earth and, therefore, humans were no longer the center of the universe, humans could still believe that they were spiritually unique and significant. Humans alone of God's creatures were moral beings, masters of a world that had apparently been designed by the Creator to support them. Humans represented the peak in the "Great Chain of Being" that united all things into a natural hierarchy. Surely no purely natural process could have led to the orderly system of life, especially to have created a thinking, reasoning, but most important a spiritual, moral being. Thus, while Copernicanism profoundly challenged man's place in the universe, his position as a unique spiritual being was left intact.

Furthermore, scientific inquiry seemed to affirm that the universe had a plan, an overall design. Newton had provided the image of a clocklike universe, being put in motion by natural laws. His single law of universal gravitation explained the orbiting of the planets and why a stone falls to the earth, but natural law was merely another way of referring to God's law. Discovering these natural laws guided thinking throughout the eighteenth and nineteenth centuries. That all events were linked together by uninterrupted cause and effect was one of the most fundamental ideas to emerge in the modern age. Indeed, scientists' faith in the universality of physical law gave analogical support for the moralist's faith in the universality of moral law. In addition, the hand of God could be seen everywhere in the laws of nature. Thus, the founders of modern science did not regard their findings as undermining traditional Christian belief. Rather, they were unveiling the means by which God fashioned His magnificent creation.[11]

Nevertheless, the new cosmology and new scientific outlook ultimately weakened traditional Christianity. Our place in the universe was being compromised with another string of developments that were occurring over a long period of time, finally culminating in Darwin's theory of evolution. In the Enlightenment, the desire to explain the universe in mechanical, materialistic terms came to the forefront. God had so exquisitely designed the universe that there was no need for Him to intervene in the form of miracles. Furthermore, miracles were contrary to what scientific findings were revealing about the regularity of nature. As natural philosophers continued their search for God's laws, the role of God faded into the background. It became increasingly less important to reconcile new findings with the Bible. The most dramatic example of this incompatibility was the discovery of fossils and the challenge they presented to the story of Genesis.

Several attempts had been made to reconcile Genesis with geology. A great deal of first-rate research in paleontology was done specifically to garner evidence in support of natural theology. However, the fossil record clearly showed that living beings had been created at widely separated intervals, not in seven continuous days. Fossils were problematic for other reasons as well. First, finding sea creatures embedded in particular strata implied that areas that were dry at one time were under water. Thus, the earth was not fixed and static, but continually changing. Second, remains were being found that did not resemble any known living organism. What were these creatures; what had happened to them? Would God have created creatures only to let them die out? It was far better to admit that the Bible contained dated and false science than try to preserve its literal meaning by harmonizing it with the latest theories in Scriptural Geology, which were always in retreat as more and more scientific knowledge came in. Findings in science should be used to help interpret the Bible rather than be twisted to be compatible with it.

In 1802 a book appeared that later was to have a profound influence on the young Charles Darwin—William Paley's *Natural Theology*. Written by an Anglican priest, the treatise was a powerful response to the threat of atheism posed by the Enlightenment thinkers. Nature gave overwhelming evidence for the unity of God that was seen in the "uniformity of plan observable in the universe."[12] In his famous opening passage, Paley drew an analogy between the workings of nature and the workings of a watch. No one would believe that a watch, with its exquisite design, the detailed workings of its springs and gears all intricately fitted together, could have come about by a natural process. The existence of the watch implied the existence of a watchmaker. However, even the simplest organism was far more complex than the most complex watch. Thus, organisms, like the watch, could not have come about by a purely natural process, but rather must be the product of a Divine watchmaker. The core of Paley's argument centered on adaptation. Example after example illustrated the remarkable adaptation of organisms to their environment, with every part of every organism designed for its function. The eye was an exquisite organ designed for sight. The human epiglottis was so perfectly designed that no alderman had ever choked at his feast. Even some species of insects were designed to look like dung to protect them from being eaten. The handiwork and divine benevolence of the Master Craftsman could be seen everywhere. God protected and looked after all his creatures, great and small.

Natural theology initially provided a powerful antidote to the findings in science that were undermining a belief in a Creator, but eventually the observations were interpreted in very different ways. Thomas Paine took the fundamental idea of the natural theologians and pushed it to its ultimate conclusion. In *The Age of Reason,* written in three parts between 1794 and 1802, he argued that the Bible had been written by men, but Nature was the handiwork of God. The Bible had been corrupted by errors in copying and translation, while Nature had an indestructible perfection. The Bible portrayed a God that was passionate and, therefore, changeable and at times vindictive, but Nature revealed Him as immutable and benevolent. Biblical revelation had come late and was supposedly revealed to one nation only, but the revelation from Nature had always been available. God communicated with humankind through magic, but Nature communicated through ordinary senses. Thus, "the theology that is now studied . . . is the study of human opinions and of human fancies concerning God." Christianity was guilty of abandoning "the original and beautiful system of theology, like a beautiful innocent, to distress and reproach, to make room for the hag of superstition." But "natural philosophy is the study of the power and wisdom of God in his works, and is the true theology."[13]

Like Paley, Paine was absolutely sure that Nature revealed a benevolent Creator. Nevertheless, relying on Nature for signs of God's benevolence was problematic. Nature was often cruel. Where was God in the slaughter that went on every day for survival? In the wild, it was eat or be eaten. An example that particularly vexed nineteenth-century theologians was the ichneumon flies. Actually a group of wasps, many species followed a perversely cruel lifestyle. Although they were free-living adults, in the larval stages they were parasites feeding on other animals, usually caterpillars, but sometimes spiders or aphids as well. The adult female pierced the host and deposited her eggs within the caterpillar. When the egg hatched, the larvae started eating from the inside. However, if the caterpillar died, it would immediately start to decay and be of no use to the larvae. Thus, the larvae ate the fat bodies and digestive organs first, keeping the caterpillar alive by preserving intact the essential heart and central nervous system. Finally, it killed its victim leaving behind the caterpillar's empty shell. Where was God's benevolence in this grisly tale?

In spite of such problems, however, natural theology offered a powerful alternative for people who were finding it increasingly difficult to continue to believe in the tenets of traditional Christianity. Instead of accepting the "whimsical account of creation—strange story of Eve—the ambiguous idea of a man/god—, and the Christian system of arithmetic, that three are one and one are three," Paine maintained that the scientific study of the structure of the universe would reveal the true power and wisdom of God.[14] Paine claimed that *The Age of Reason* was a tract in support of Deism, but people perceived it primarily as an attack on all organized religion, not just Christianity. He was attacked as an atheist and died a pauper. Nevertheless, Paine's ideas resonated with the thinking of a small, but influential group of theologians who were determined to make the findings of science compatible with religious belief in the Deity. Their work became known as the higher criticism.

The convergence of scientific naturalism and historical criticism came to a head with the publication of *Essays and Reviews* in 1860. Just as Galileo had argued two hundred years earlier that science and theology were two separate spheres of knowledge, the authors, a group of liberal Anglican clergy, claimed that the Bible was not in the business of interpreting nature, nor should it be. Galileo believed the Bible could never speak untruth whenever its true meaning was understood, but the Bible could be difficult to understand. Like Galileo, the authors urged that reason and evidence from all possible sources should be used to interpret and understand the Bible. Mosaic cosmogony was not an authentic utterance of Divine knowledge, but a human one. It was to be used in a special way for educating humankind. At one time the gospel miracles had constituted proof of Christ's divinity, but now they were an embarrassment and a liability. For the faithful, this book was more of a shock than Darwin's *The Origin of Species,* published the year before. Darwinism was threatening for many reasons, but in the religious community it was because it was part of a much wider revolution in attitudes toward the Bible. *Essays and Reviews* was condemned by the bishops of the Anglican Church, with the Archbishop of Canterbury issuing an encyclical against it.

Biblical criticism was alive and well on the continent as well. In Germany, David Frederich Strauss painstakingly analyzed the Gospels, pointing out all their inconsistencies and argued that trying to harmonize them was impossible. For him, many of the healings performed by Christ as well as the Resurrection were not supported by a careful historical approach and were simply untrue. Far better to acknowledge that the accounts of the original incidents had been altered over time, due to differing interpretations and embellished to the point of legend. Strauss argued that to continue to cling to an outmoded literalism would only bring Christianity into disrepute. He hoped to preserve the moral insights of Christian doctrine by making them independent of miracles that a modern educated person could no longer believe.

The idea that nature does nothing in vain was an ancient theme. However, the argument from design both furthered the belief in a Divine creator and undermined Christianity at the same time.[15] Naturalism had great appeal to both religious and nonreligious thinkers alike. Many religious thinkers, in fact, welcomed the developments in science, regarding them as an aid to faith. Natural theology could serve as a mediator between different theological positions by offering independent proof of a God who had also revealed Himself in the person of Christ. Deists also liked the design argument, because the more that could be known of God through rational inference, then the less dependent one would be on revelation and miracles to justify their faith. In addition, many people welcomed the challenges to orthodox religious belief and felt tremendous relief, no longer burdened by the Christian doctrine of original sin.

Natural theology was also compatible with a metaphysical conception of the universe that was finding increasing support as a result of findings about the evolution of the universe and life on earth, where evolution in this context simply means change. By 1850, the evidence from paleontology suggested that the history of life showed a great progressive development, from amoebas to fish to reptiles to birds to mammals, finally culminating in the coming of humankind. The *Vestiges of the Natural*

History of Creation specifically argued that species changed in accordance with a law of organic development resulting in increasingly higher and more complex organisms. Natural theology argued that the order and complexity of the world, particularly as exemplified by living organisms, could not have come about by purely naturalistic means, but was the result of an intelligent designer. Chambers used the argument from design, but couched it in evolutionary terms. *Vestiges* even suggested that species superior to us have not yet evolved. Evolution within species would continue and humankind might continue toward higher and nobler developments. In contrast to a faith in science where progress depended on the applications of scientific methods to societal problems, this was a faith in progress as a law of the universe.[16] Such a process would occur worldwide, independently of human efforts. In 1859 Darwin reinforced this idea writing in *The Origin* that the process of natural selection resulted in organisms becoming increasingly well adapted to their environment, which also led to an overall improvement.[17] For those so inclined this suggested that a Divine Craftsman was overseeing the history of life.

Nevertheless, Darwin had turned Paley's argument on its head. For Paley, the relationship between structure and function that resulted in the remarkable adaptations of organisms was powerful evidence for supernatural design. For Darwin, however, adaptation became a natural process by which organisms adjust to a changing environment. There was no need for an intelligent designer. Scientific findings in many other areas also suggested that we live in a materialistic world, devoid of spiritual meaning. In this godless world many Victorians turned to the "pseudosciences" for solace. This pejorative label usually appeared in hindsight or was employed as a way to discredit a particular line of inquiry that seemed threatening to the status quo. However, in the Victorian era the pseudosciences, or what I prefer as a more accurate description, the marginal sciences represented an attempt to come to terms with the most critical problems in science, philosophy, and religion. Although many Victorians were willing to give up specific Christian doctrine such as the literal interpretation of the Bible, very few people were prepared to become atheists. Recent findings in physics and chemistry such as the discovery of electromagnetism gave hope that it might finally be possible to scientifically demonstrate the existence of the spirit world.

Chapter 4 explores the rise of the modern-day spiritualist movement and examines in detail William Crookes's investigation of two prominent mediums, Douglas Daniel Home and Florence Cook along with the fully materialized spirit of "Katie King." Crookes, a chemist and physicist, was highly skeptical of spiritualist claims and promised to use his outstanding experimental skills in a rigorous examination of spiritualism. However, as in the previous case studies, Crookes's involvement with spiritualism is a multilayered story that reveals many factors shaped his investigation, and also points out the less savory aspects of the spiritualist movement. In the public arena spiritualism became a haven for hucksters and quacks who preyed on individuals grief stricken by the loss of loved ones as well as exploiting the deep-seated desire among Victorians to reconcile the findings of science with religion. Although Crookes became thoroughly disenchanted with spiritualism, he remained convinced that some undiscovered psychic force existed, and he was not alone in that regard.

Spiritualism attracted serious interest among a small, but significant group of scientists, particularly physicists who wanted to understand psychic phenomena. Nevertheless, spiritualist investigations never became part of mainstream science, but it was not just because most of the phenomenon were exposed as fraudulent. The Society for Psychical Research investigated all kinds of unusual mental phenomena and produced high quality reports. However, they were largely unsuccessful in connecting with more orthodox branches of psychological research. Spiritualist phenomena were illusive, and erratic. Some of the most distinguished physicists of the day argued that physical laws were uniform and universal. They reiterated the importance of repeatability to good experimental design and claimed that spiritualism violated the basic principles of good scientific practice. Their views prevailed in defining not only what counted as good scientific practice, but also in limiting the kinds of topics deemed legitimate in their quest for the professionalization of science. However, not all scientists were convinced that spiritualist phenomena were outside the purview of scientific investigation. An undiscovered force could potentially explain psychic phenomena without recourse to spirit agents. In addition, demonstrating that mind was not just an epiphenomenon of the brain was critical to spiritualists and psychic researchers in their argument against materialism.

Many spiritualists also found Darwin's theory very attractive, interpreting it in such a way as to meet their spiritual needs. Chapter 5 explores the intersection of spiritualism with evolutionary theory, focusing on the career of Alfred Russel Wallace. For traditional Christians evolutionary theory was problematic. If humans had evolved from a lower form, what were the implications for the immortality of the soul and the uniqueness of human beings? However, most British spiritualists including Christian spiritualists with their limited knowledge of biology happily embraced evolutionary theory because for them it confirmed their belief that higher forms would continue to evolve in the spiritual realm after physical death.

Wallace independently came up with the theory of natural selection and in many ways was more of a selectionist than Darwin. He understood only too well the implications of evolutionary theory and that it challenged the most basic tenets of spiritualism. Yet he managed to reconcile his belief in evolution by natural selection with an absolute faith in spiritualism. He did not think natural selection was wrong, only incomplete, claiming that neither natural selection nor a general theory of evolution could give an adequate account of consciousness and a variety of other human traits. He made no reference to spiritualist ideas, basing his objections on grounds of utility only.

Wallace raised a number of serious issues surrounding human evolution that needed to be addressed and the Darwinian camp responded. Although Thomas Huxley had argued passionately that the probable lowly ancestry of humankind did not detract from their unique status in the panoply of life in *Man's Place in Nature* in 1863, his defense of Darwin did not address one of Wallace's key objections. Wallace had argued that there would be no need for humans living in their primitive state to have a big brain and particularly a high sense of morality. In *The Descent of Man* Darwin explained how the evolution of the moral sense could have come about. Thus Wallace's objections helped further the discussion over human origins. A variety of experiences shaped his mature views including his extended period of living with native peoples, which made him more open to spiritualism and contributed to his doubts concerning natural section as a complete explanation for the emergence of humans in the history of life.

Wallace's career, like Crookes's, illustrates that it would be a mistake to consider their investigations into spiritualism separately from their more mainstream scientific research. Both men were not arguing in favor of the supernatural, but rather of extending the boundaries of what should be considered science. Many spiritualists and not a few scientists had a vision of a "new science," which would unify spirit and matter, mind and body. If such a vision could be achieved, it would reconcile the growing gulf between science and faith.

These chapters illustrate again and again that the scientific community did not speak in one unified authoritative voice concerning what was considered legitimate scientific knowledge. "Issues of place, practice, and audience have been central to the construction of scientific authority and orthodoxy."[18] Chapter 6 explores some of the negotiations surrounding evolutionary theory by examining the ideas of Thomas Huxley, and in doing so elucidates why evolution did not share the same fate as the previous case studies. Like phrenology, evolutionary ideas pre-Darwin had found their greatest support among the dissidents and radicals of the London medical schools.[19] Just like the ideas promulgated in *Vestiges of the Natural History of Creation*, including its phrenological content, evolution was attacked on both scientific and religious grounds. It should be pointed out that phrenology was not initially attacked because it was stupid, but rather that it was a materialistic doctrine. But unlike phrenology, evolution did not remain marginalized. The scientific naturalists eventually gained control of the universities and also largely succeeded in defining scientific methodology and what were legitimate topics of scientific investigation.

Huxley played a leading role in defending evolutionary theory against religious attacks. He also used Darwin's theory to promote the scientific and professional status of biology. However, Huxley was skeptical of the two basic tenets of Darwin's theory, natural selection and gradualism. Huxley's doubts are the basis for exploring some of the scientific objections to Darwin's theory. They also provide a window to view the emerging structure of what becomes defined as good scientific practice and suggest why unlike the previous examples, evolutionary theory emerged as a powerful and unifying theory for the life sciences.

Based on the evidence from embryology and comparative anatomy of both living and extinct organisms, and the pattern of the fossil record, Huxley thought species were fixed. Organisms appeared to be grouped into distinct types with no transitions between them. However, Darwin's interpretation resulted in Huxley recognizing that the evidence was also compatible with species descent from a common ancestor. Nevertheless, he still maintained that the fossil record did not support the idea of gradual change. Saltation allowed Huxley to keep his belief in the idea of distinct types, explain the gaps in the fossil record and accept evolution. As more and more transitional fossils were discovered and new findings from development suggested the common ancestry of all organisms, Huxley eventually accepted the idea of slow gradual change. However, he continued to view natural selection as a hypothesis rather than a proven theory or fact.

While Huxley thought natural selection played an important role in the history of life, he maintained that the evidence did not yet exist that demonstrated it had the power to create good physiological species incapable of interbreeding, and not merely well marked varieties. He agreed with Darwin that it could be difficult to determine what was a variety and what was a species. He recognized that breeding experiments often gave inconclusive results. Darwin in turn also recognized that if he could demonstrate that hybrid sterility was actually selected for, his theory would be in a much stronger position. Because of Huxley's objections, Darwin performed a series of experiments that he might not have and eventually strengthened his original position that hybrid sterility was not specifically selected for, but was an incidental byproduct of other selection. Huxley remained unconvinced.

Huxley's disagreement with Darwin highlights what is the fundamental question of my study. How is evidence interpreted and evaluated, and what kinds of factors influence that judgment? Geologists based on very fragmented evidence posited an elaborate scenario concerning the history of life. In the early part of the nineteenth century was the quality of their evidence and their interpretation of it that different from that of some of the scientists who were investigating the phenomena observed at séances? Darwin used the same kinds of evidence to support the evolution of the moral sense as Gall did in support of phrenology. Yet spiritualism and phrenology were eventually discredited. Huxley's concerns over natural selection help elucidate how these judgments were made. He did not disagree with the interpretation of the experimental results, but rather over what constituted proof of a hypothesis.

Huxley was following William Herschel, maintaining that theories should be based on direct empirical evidence. In making his case for natural selection, Darwin was following William Whewell, claiming that if a theory had wide explanatory power than it should be accepted. In *The Origin*, however, Darwin had made use of both types of arguments, both empirical and the idea of consilience. Huxley also recognized the importance of consilience claiming that no theory except for evolution could explain so many "facts" of the natural world. Yet he maintained that natural selection had to be experimentally demonstrated to move it from hypothesis to theory. Darwin felt that Huxley was placing an essentially impossible demand on his theory. This discussion also highlights that the distinction between the two different approaches is not absolute. The influence a theory ultimately wields depends on how well it makes sense of empirical data. Both are essential aspects of good scientific methodology.

Huxley avidly defended Darwin's theory in spite of his disagreements with it because he recognized that it was extremely powerful as a means to further the understanding of the natural world. Initially, he did not think either transmutation or natural selection had been proven, but these ideas could be tested. Evolution did not suffer the same fate as the other case studies in this book, not because it turned out to be correct, but rather because its advocates not only promoted, but actually practiced for the most part the scientific methodology that they espoused. The quality and quantity of evidence in favor of evolution was simply much higher than in the previous case studies. One of the hallmarks of a good scientific theory is that it should continue to generate hypotheses that can be tested. Evolutionary theory has proved to be exceeding robust in this regard. The very issues that Huxley was concerned with, such as the role of development and the gaps in the fossil record, have continued to be revisited. Disagreements and disputes among evolutionists are the marks of a good scientific theory. Nevertheless, evolution is not immune from the same kinds of problems that have plagued the marginal sciences.

The final chapter examines the culture of science and why it has been so successful in providing reliable knowledge of the natural world.

Philosophers have long been grappling with how to differentiate science from other types of knowledge, but none of their explanations have been entirely satisfactory.[20] Henry Bauer has suggested that the notion of a filter represents a more accurate description of how scientific knowledge accumulates and is particularly applicable to my case studies.[21] In early stages of an investigation knowledge is highly unreliable, coming from many different sources, but various filtering processes winnow the possibilities of what will eventually become accepted as scientific knowledge. In a later stage, part of the filtering process is the actual creation of boundaries. However, the crucial aspect of the filtering process that determines what eventually becomes reliable scientific information in the long term is that science is fundamentally a cooperative enterprise. Theories and "facts" must survive a period of testing and experimenting. Darwin's theory survived this period in a way that the other case studies did not, while at the same time remaining controversial.

`Examining two present-day controversies that connect to the case studies illustrates that the boundaries in science are ever changing as scientists continue to expand into areas that were once the domain of religion and philosophy and as new areas of investigation continue to vie for scientific legitimacy. The first explores the role that fossils continue to play as possible links between myth and science. The second examines evolutionary psychology, which claims to provide a definitive scientific explanation of human nature, just as phrenology did in times past. Both of these investigations have spawned contentious debates. As with the early stages of the case studies in the book, it is unclear what their eventual status will be. Since human beings with all their foibles, biases as well as their keen intellect produce scientific knowledge, such knowledge will never be totally objective. However, it is precisely these attributes that contribute to the multiplicity and productivity of scientific ideas. In the short term, in what Bauer refers to as frontier science, many factors come into play that determine what kinds of questions are asked, what counts as evidence, and how evidence is interpreted. However, science in the long term is self-correcting. Scientific method is important and evidence does matter. As all these case studies demonstrate, scientists provide us with an ever changing, but increasingly accurate view of the material world.

2

Swimming at the Edges of Scientific Respectability

Sea Serpents, Charles Lyell, and the Professionalization of Geology

. . . and what we took for a shoal of black fish was nothing less than the bunches on the back of the celebrated SEA SERPENT!

—Timothy Hodgkins, "August 15, 1818"

A larger body of evidence from eyewitnesses might be got together in proof of ghosts than of the sea serpent.

—Richard Owen, "The Great Sea Serpent"

I can no longer doubt the existence of some large marine reptile allied to Ichthyosaurus and Plesiosaurus yet unknown to naturalists.

—Louis Agassiz, "1849 Lecture"

IN 1819, PEOPLE in Boston clamored to the theater to see "The Sea Serpent or Gloucester Hoax." As the drama unfolded, fishermen told tales of a terrible serpent scaring off the fish, and a young man schemed to capture the serpent to earn a reward that would allow him to marry his beloved. The play was a variant on a theme that has entertained humans for eons: man fends off horrible (often mythical) monster in order to win the hand of the maiden he loves. But plays are not merely entertainment;

rather, they reflect the issues of the day, whether they are political, social, religious or scientific. In recent years there had been several sightings of a sea serpent off the New England coast and the story line provided the dramatic vehicle to explore a question that was on the minds of an increasing number of people on both sides of the Atlantic. Were sea serpents a topic worthy of scientific investigation, or as the title claimed merely a hoax?

Not wanting to burden his audience with undue subtleties, the author used a character named Scepticus to be the voice of doubt. "The World's turned typsy turvy and mankind/ in quest of monsters monstrously incline/ rack their invention prodigies to find." Linnaeus (whose name, as will be made clear later, was also very deliberate) disagreed, telling the unbeliever:

> evolving time
> as he unrolls the destinies sublime
> Of nations, insects fishes, birds, and men
> bids no the sameself scenes recur again
> But with new characters adorns the page
> and gives distinctness to each passing age.
> There is no genius where there's nothing new
> And nothing wonderful in what is true.[1]

Scepticus, ever cynical, responded that "blockheads are always ready to believe and there will always be knaves eager to deceive." However, Linnaeus countered that "Philosophers do more, With artificial optics they explore, Earth, air and ocean—all there is to see." Indeed, it was precisely the developments in science that were responsible for a revival of interest in sea serpents. Increased exploration of the sea combined with paleontological finds suggested that the "mythical" sea monster might not be a myth after all.

In the south of England in the secondary strata known as the Lias by local quarrymen, crocodile-like fossils were discovered that were unlike any known organism. One had paddles rather than legs and paleontologist William Conybeare named it *Ichthyosaurus* (fish-lizard) to emphasize the ambiguity of its classification. Even stranger was the *Plesiosaurus* (almost-lizard), so named because Conybeare thought it was anatomically intermediate between the *Ichthyosaurus* and the crocodile, writing that it looked "like a snake threaded through the body of a turtle."[2] Real evidence of monsters that walked the earth and swam in the seas of ancient times brought a renewed curiosity about the existence of present-day monsters.

Several prominent scientists went on record in support of the serpent's existence. In a lecture in Philadelphia in 1849, Louis Agassiz told his audience that the general character of the continent, as well as the fact that other Paleozoic types were still alive,[3] led him to conclude that a living representative of these ancient marine reptiles would soon be found. Henrich Rathke wrote, "There can clearly be no doubt that in the seas around Norway there is a large serpentine animal which may reach considerable length."[4] Sir William Hooker claimed the serpent:

> can now no longer be considered in association with hydras and mermaids for there has been nothing said with regard to it inconsistent with reason. It may at least be assumed as a sober fact in Natural History quite unconnected with the gigantic exploits of the God Thor or the fanciful absurdities of the Scandinavian mythology. We cannot suppose, that the most ultra-skeptical can now continue to doubt with regard to facts attested by such highly respectable witnesses.[5]

Charles Lyell, the foremost geologist of the English world, believed in the sea monster's existence, recognizing it would provide dramatic evidence in favor of the most controversial aspect of his theory of life's history.

It was not mere coincidence that as paleontologists began dredging up plesiosaurs and other relics from the past that a dramatic increase in sightings of sea serpents also occurred. How seriously should one regard these eyewitness accounts? As the editor of a collection of articles and scientific reports of sightings asked, "What has become of the Sea Serpent?" Invoking Darwin's theory of evolution, he pondered, "Was it the last of a species of ancient reptiles and has gone the way of all extinct creatures like the dinosaur . . . or was he the creation of [sailor] Jack's lurid imagination, perhaps influenced by his sampling the cargoes of heady West Indie rum while on the high seas?"[6] In fact, the vast majority of serpent sightings were not from drunken sailors, and the resulting discussion that attempted to link the sightings with the fantastic creatures of the fossil record resulted in the sea serpent achieving a level of scientific legitimacy in the nineteenth century that it never had before or since.

Nevertheless, sea serpent investigations remained at the margins of scientific respectability for a variety of reasons. As geology struggled to achieve professional status the serpent became caught up in boundary disputes as both the sightings and the interpretation of them were questioned. The controversy over the serpent's existence reflected differing views of earth history and on the role of amateurs and professionals in contributing to natural knowledge. With a history inextricably intertwined with mythology the serpent would be at a distinct disadvantage in capturing the attention of an emerging class of professional scientists who wanted distinct boundaries drawn between scientific and other kinds of knowledge.

The Status of Sea Serpents before the Nineteenth Century

The sea serpent brings to mind most often a creature of fantasy and legend, a gigantic monster that struck terror in the hearts and minds of all who came into contact with it either personally or through hearing about such encounters. From time immemorial, seafarers have brought back tales of monstrous serpents who destroyed large ships, swallowed shoals of fish, or even an entire boat of men in one giant gulp.[7] One of the earliest accounts of a sea monster is from Homer. In the *Odyssey* Ulysses must sail through the narrow strait between the many-headed monster Scylla ("she has twelve splay feet and six lank scrawny necks. Each neck bears an obscene head, toothy with three rows of thick set crowded fangs blackly charged with death") or the deadly whirlpool of Charybdis. Scylla was perhaps an octopus or squid, but with embellishments befitting a Homeric epic. Aristotle also described what probably was a giant squid; and from the accuracy of his description, it is clear that he had an actual specimen. However, he also depicted animals from the sea that could not be classified by kind because of their rarity. "Among fishermen with long experience some claim to have seen in the sea animals like beams of wood, black and round and the same thickness throughout. Other of these animals are like shields: they are said to be red in colour and to have many fins."[8] While the latter sounds like the giant squid, was the former a description of a sea serpent?

Pliny described a giant polyp with thirty-foot arms and claimed it climbed out of the sea to eat fish that were being salted at Carteia in Spain. In the Middle Ages, a fantastic creature that was a combination of a giant squid and a whale gave rise to the legend of the island beast. It was described in one of the earliest bestiaries, the *Physiologus,* and reached its peak of popularity in the legend of St. Brandon. This island beast was still taken seriously in 1555 in Olaus Magnus's *Historia de Gentibus Septentrionalibus* (History of the people of the Northern Regions).

Magnus was the Catholic archbishop of Sweden and an influential cleric and historian. However, he is most remembered for the *Historia,* which was filled with detailed descriptions and richly illustrated with fantastic creatures that were copied, reproduced, and modified for centuries. He described several different sea monsters including a gigantic fish over 220 feet long with huge eyes and claimed, "one of the Sea Monsters will drown easily many great ships provided with many strong Mariners."[9] Another sea serpent over two hundred feet long and twenty feet in diameter lived in the rocks and holes near the shore of Bergen. "It comes out of its cavern only on summer nights and in fine weather to destroy calves, lambs, or hogs, or goes into the sea to eat cuttles, lobster, and all kinds of sea crabs. It has a growth of hairs of two feet in length hanging from the neck, sharp scales of a dark brown color, and brilliant flaming eyes."[10] A woodcut entitled "Les Marins Monstres & Terrestres" reproduced many of Magnus's illustrations, but also added many new creatures. They were included in Conrad Gesner's massive 4,500-page treatise *Historia Animalicum* (1551–1558), which is considered the basis for modern zoological classification. It contained everything that was known, imagined, and reported about all animals from the time of Aristotle. The book was a mixture of fact and fantasy, but how to sort out which was which was not easy. As the old saying goes, "truth is stranger than fiction" and many real creatures were indeed quite fantastic.

Hans Egede, a Danish missionary who eventually became the Bishop of Greenland, described a sea monster he saw in 1734 in his *Det Gamle Grønlands nye Perlustration* (1741):

> The Monster was of so huge a Size, that coming out of the Water its Head reached as high as the Mast-Head; its Body was as bulky as the Ship, and three or four times as long. It had a long pointed Snout, and spouted like a Whale-Fish; great broad Paws, and the Body seemed covered with shell-work; its skin very rugged and uneven. The under Part of its Body was shaped like an enormous huge Serpent, and when it dived again under Water, it plunged backwards into the Sea and so raised its Tail aloft, which seemed a whole Ship's Length distant from the bulkiest part of its Body.[11]

By this time, both the Dutch and British had a thriving whaling industry; and therefore, we can presume that Egede was familiar with whales. It is unlikely that he would have mistaken what was a whale for this creature. Furthermore, he claimed that he had not seen any other type of sea monster that had been described in earlier accounts. Egede was described as a sober and trustworthy man of the cloth, and he had Pastor Bing draw a picture of the monster that became one of the earliest illustrations of a sea monster based on a "reliable" eyewitness account.

Many "mythical" creatures were undoubtedly exaggerated descriptions of real animals such as the giant squid and octopus. The best example of this phenomenon was the Kraken of Scandinavia. Erik Ludvigesen Pontoppidan, the Bishop of Bergen, provided the most detailed description of this creature in *The Natural History of Norway* (1755). The bishop gave a quite plausible description of a giant squid and its behavior except that he also wrote that the whole body of an adult had probably never been seen, and he estimated that "its back or upper part seems to be in appearance about an English mile and a half in circumference."[12] However, not all sea monsters can be accounted for so easily. The *Natural History of Norway* was filled with descriptions of creatures that did not seem to be like any known organism. After taking a deposition from Captain von Ferry on a sighting, the bishop requested that the captain write a letter to the court justice at Bergen describing the "sea-snake" he had seen passing his ship in August 1746:

> The head of this sea serpent, which it held more than two feet above the water, resembled that of a horse. It was of greyish color, and the mouth was quite black and very large. It had large black eyes, and a long white mane, which hung down to the surface of the water. Besides the head and neck, we saw seven or eight folds, or coils, of this snake, which were very thick, and as far as we could tell, there was a fathom's distance between each fold.[13]

Throughout history serpentlike creatures have been described that did not fall into any known classification. Thus, sea monsters defied easy categorization and transgressed the boundaries between the natural and supernatural, the real and the imaginary.

Nevertheless, until about 1750 the sea serpent was rarely heard of outside of Scandinavia. Pontoppidan and his countrymen may have believed in the reality of the sea serpent in the form of the Kraken, but the rest of the Western world relegated it to the category of myth and legend along with mermaids and unicorns. Many other kinds of monsters had been classified, no longer associated with a supernatural origin. By the nineteenth century, the investigation of genuine anomalies such as extra limbs had been incorporated into embryological studies and was recognized as the subdiscipline teratology. However, the boundaries between the real and the imaginary were still being negotiated. Mermaids had been the object of so much hucksterism and sensationalism that mermaid sightings were never really taken as serious evidence. Only actual specimens counted and these were invariably discredited. But this was not true for the serpent sightings. Sea serpents had never been captured nor exposed as fakes. Since the sea was still relatively unexplored, it seemed entirely plausible that strange creatures might inhabit its depths.

In the nineteenth century the number of exploratory voyages increased dramatically. Their purpose was not just to chart the seas more accurately, but also to specifically collect fauna and flora from all over the world. A cornucopia of exotic organisms never seen before was brought back to be dissected and classified in an attempt to comprehend the natural history of the planet. And many of these organisms were indeed strange. In 1799, when first describing the platypus George Shaw wrote:

> It seems most extraordinary in its conformation, exhibiting the perfect resemblance of the beak of a Duck engrafted on the head of a quadruped. So accurate is the similitude, that, at first view, it naturally excites the idea of some deceptive preparation.[14]

But Shaw could find no stitches, even upon dissection. The platypus not only looked bizarre, but defied classification into any known group having characteristics of birds, mammals, and reptiles. Not only did these findings lead to a fuller understanding of the richness and diversity of life, but as fossil remains were found with increasing frequency, they also provided a glimpse into the past. It is not surprising that sea serpents were observed, swimming in this environment of exploration and discovery.

Sea Serpents and the History of Life

If the sea serpent was to be taken seriously, no longer the product of ancient storytellers' over active imaginations, but a real animal from the past, then its existence would have to be compatible with the prevailing views of earth history. However, these views were in flux, in large part due to the tremendous advances that were taking place in geology, physics, and natural history. Nevertheless, by the turn of the nineteenth century most geologists viewed earth history developmentally and argued that general overall trends through time could be detected on a global scale. This directional view of earth history was based on the geophysical theory of a cooling earth. Georges-Louis Leclerc Buffon had popularized the idea of a gradually cooling earth in his speculative *Epochs de la Nature* (1778) that had increasing support in the nineteenth century. By the 1820s the theory enjoyed considerable prestige because of several different kinds of evidence. First, the idea of a "central heat" had substantial empirical support. Louis Cordier had shown that a geothermal gradient was not only real, but also universal. In addition, the application of Fourier's physics had been used to explain both the idea of a cooling earth through time and the geothermal gradient. Fourier had demonstrated that the temperature along the earth's surface plotted against time had fallen along an exponential curve. In the recent epochs the heat loss from the earth's surface was quite small compared to the effects of solar radiation, resulting in fairly uniform and stable conditions.[15]

A theory of residual central heat implied a directionalist interpretation for a variety of phenomena and suggested that environmental conditions on the earth were very different in the ancient epochs than in more recent ones. It was not unreasonable to conclude that such different physical conditions meant the organic world also would be vastly different. At first the world was too hot for any kind of life. Reptiles dominated the Secondary period because they were suited to a tropical environment. As the earth continued to cool, mammals made their appearance in the Tertiary period. The many new fossil discoveries were providing evidence for just such a pattern.

Georges Cuvier's work clearly demonstrated that organisms had become extinct and that entirely new faunas had replaced them. As the environment, particularly climatic conditions, changed new organisms appeared that were adapted to those conditions. They flourished for a time, but became extinct and were replaced by other organisms that were better adapted to the new environment. Thus, a developmental view of earth history was based not just on geological change, but on the changes in the organic world as well. Cuvier's work along with the evidence of a cooling earth resulted in organic and geological change being linked, each providing support for the other. A directionally changing environment explained the directionally changing nature of successive faunas. Most geologists also believed that the earth had been shaped by catastrophic events, whole mountain ranges being created or destroyed virtually all at once. The theory of the cooling earth provided an explanation for why in the past events would have been more violent than in the present.

Charles Lyell in *The Principles of Geology* (1830–1833) challenged this commonly accepted view of earth history, arguing instead for what became known as uniformitarianism.[16] There were several aspects to uniformitarianism. First, it was a totally naturalistic account of the history of the earth. Lyell wanted a scientific geology and believed that the prevailing catastrophist methodology would prevent geology from rising to the level of an exact science. In the early nineteenth century, catastrophism in Britain was associated with scriptural geology, that is, findings in geology were used as evidence for the biblical flood. Although catastrophism did not require supernatural intervention, this was its legacy. Uniformitarianism was explicit in its denial of supernatural forces to explain earth history. Second, actualistic processes shaped the earth: causes now operating to alter geological formations were the same as those operating in the past. Present-day processes such as erosion, earthquakes, and volcanoes were of the same sort and intensity as those acting in the past. If processes occurring today were basically the same as those occurring in the past, actualism provided a methodology to investigate past events. Thus, uniformitarianism was more than a theory about earth history; it provided a research program for understanding the past. Third, geological processes were slow and gradual. These ideas were crucial to Darwin's own thinking about evolution and they helped pave the way for the acceptance of his theory. Uniformitarianism suggested evolution in the organic world as much as in the inorganic. The origin of a new species by supernatural agency would represent a much greater "catastrophe" than the ones that Lyell had successfully eliminated from geological speculation. There was one other aspect to uniformitarianism, and it was this component that specifically argued against a directional progressive view of earth history. It also led Lyell to become the sea serpent's most eminent advocate.

Lyell revived James Hutton's view that the history of the earth exhibited a steady-state pattern of endlessly repeatable cycles. "We find no vestige of a beginning—no prospect of an end."[17] Rather than the successive appearance of increasingly complex organisms, Lyell believed that representatives of all classes of organisms existed at all times. He claimed that the fossil record only looked progressive because of its incompleteness. Such a position was not unreasonable in the 1830s. Paleontology as a rigorous discipline was barely twenty years old. The accumulated findings yielded an extremely incomplete record and, thus, interpretation was based on a very fragmented pattern. Lyell maintained that "reading" the fossil record was like reading a book that had most of the pages missing and few words left on the remaining pages. Since sampling of the record was quite limited and displayed many anomalies, the case for progression was far from conclusive. However, as the number of fossil finds continued to mount, this position became increasingly difficult to defend. Although other aspects of his uniformitarianism were relatively well accepted, his steady-state cosmology never had many adherents. Lyell's minority viewpoint received a further challenge in 1844 with the publication of the *Vestiges of the Natural History of Creation. Vestiges* argued not just for transmutation, but clearly portrayed the fossil record as progressive, with more complex organisms appearing through time.

In trying to make his case for a steady-state worldview Lyell had to argue repeatedly that negative evidence should be discounted. He maintained that just because certain organisms have not been found in the older strata does not mean that we should conclude they do not exist. In addition, new fossil discoveries continued to push the origin dates of specific groups back earlier and earlier in time. Lyell used these findings to argue that groups that appeared late in the fossil record such as mammals would eventually be discovered in the early epochs. Nevertheless, as more and more fossils came in, the overall scheme remained progressive. Vertebrates appeared later than invertebrates, reptiles later than fish, and mammals simply were not turning up in the early Silurian epoch (see figure 2.1 for classification of the periods of life history). His antiprogressive position was becoming less and less likely to be true. However, there was another line of argument that Lyell could use. Rather than trying to find fossil evidence of modern species, what if he could find evidence of ancient organisms existing in the present? A sea serpent would be just such evidence. Lyell began to collect documentation of sea serpent sightings.

In North America, there had been a few scattered sightings of the sea serpent between 1800 and 1810. They were dismissed as cases of mistaken identity and essentially ignored, but in 1817 a great rash of sightings occurred. Too many people had observed a serpent on several different occasions, and the Linnaean Society appointed a committee to

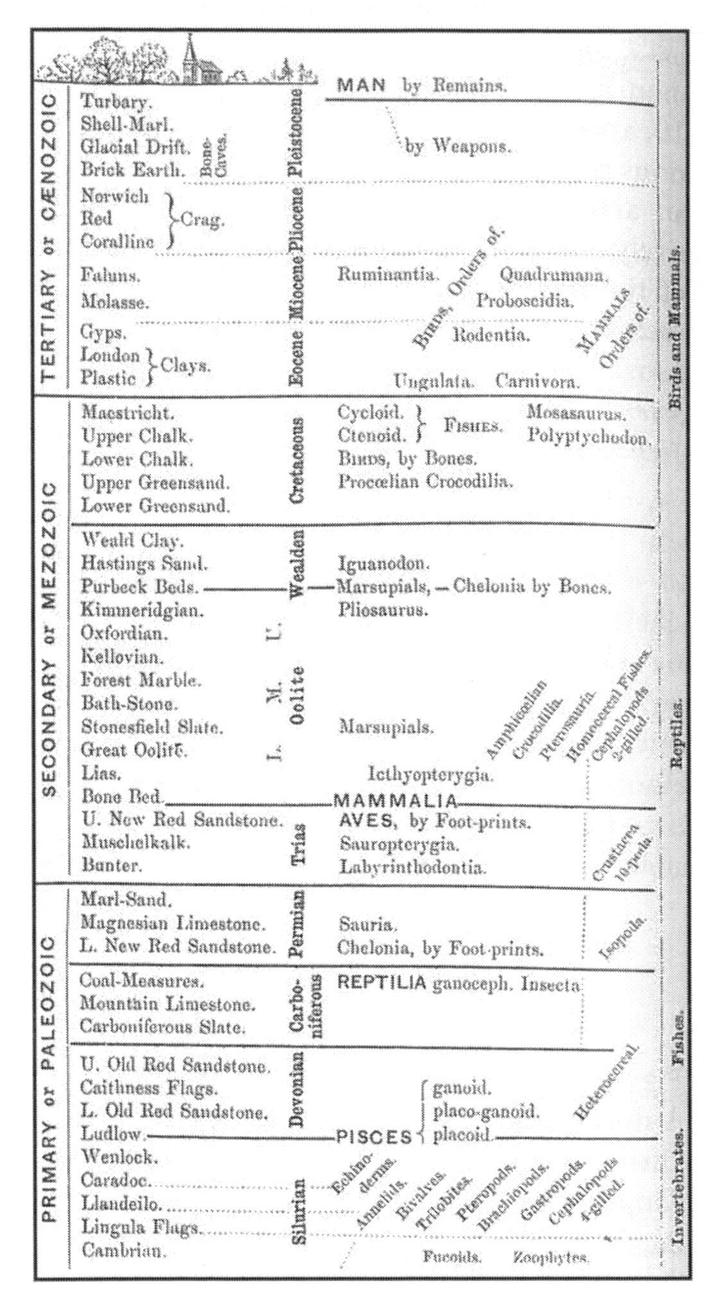

FIGURE 2.1. Table of Geological Strata and First Recorded Appearances of Fossil Animals. From Richard Owen, *Palaeontology* (Edinburgh: Black, 1861). Published in Sherrie Lynne Lyons, *Thomas Henry Huxley: The Evolution of a Scientist* (New York: Prometheus Books, 1999). Reprinted by permission of the publisher.

collect any evidence of the creature. Witnesses were asked to provide written testimony describing what they saw, which was then signed and sworn before a magistrate.[18] Mathew Gaffney of Gloucester, a ship's carpenter, gave one such affidavit:

> On the Fourteenth day of August A.D. 1817 between the hours of four and five o'clock in the afternoon, I saw a strange marine animal resembling a serpent in the harbour in said Gloucester. I was in a boat and was within thirty feet of him. His head appeared full as large as a four-gallon keg, his body as large as a barrel, and his length that I saw, I should judge forty feet at least. The top of his head was of dark colour, and the under part of his head appeared nearly white, as did also several feet of his belly. . . . His motion was vertical like a caterpillar.[19]

In addition, a detailed questionnaire was developed that witnesses were asked to answer. It was no coincidence that the character in the play that defended the possibility of the sea serpent's existence was called Linnaeus.

There continued to be several sightings per year off the coast of New England for the next several years that were reported in papers all over the country. The *Niagara Patriot* wrote that the serpent had been seen again in the harbor of Gloucester, and this time allowed for much more favorable viewing than in the past. Mr. W. Sergent along with two young men were fishing and observed what they thought was a hake approaching. It came within a few feet of Mr. Sergent's boat with its head about a foot above the water, and passed under the boat with a gliding motion, "apparently without any muscular exertion." Mr. Sergent described the head as resembling:

> that of a toad, having projections which he calls smellers, like the hake, on each side. On top of the head were three spears or horns, the middle one the largest, and all three at that time laying flat on the head. The body appeared about the bigness of a lime cask. The color was a dark brown. . . . The scales about the size of the crown of his hat, and the barnacle on the body about four inches.

He estimated the creature to be about seventy to one hundred feet long. He attempted to obtain a barnacle from the serpent's back, but when he struck the water with an oar, "the Serpent sunk beneath the surface without any struggle and disappeared."[20] Once there had been a sighting, inevitably many other reports followed, but they were not all accepted uncritically. Two weeks after Mr. Sergent's account was printed, a Captain Smith also reported sighting what he thought was the serpent. But later in the day, he saw nearby at two different times what he unequivocally identified as a whale, and concluded that he had been mistaken in claiming he had seen the serpent.[21] Nevertheless, the hunt was on. Convinced that the serpent existed, a Captain Rich specifically went searching for it, but was unsuccessful. Although he regretted disappointing the public, he was confident that his failure to capture the serpent "cannot invalidate the mass of evidence which has been adduced by hundreds of witnesses in favor of the appearance of such an animal on our coast."[22] In September of 1822, there were sightings virtually every day "by different individuals of the highest respectability."[23] As we will see, however, who qualified as a reliable witness was a crucial aspect of the sea serpent controversy.

In addition, the descriptions of the serpent varied considerably. The ship *Douglas* described a creature:

> The height out of the water was about 10–12 feet, with flippers like a turtle on each side, a third of the way from the tail. Length of the flippers from 12–15 feet, one on each side, near the tail 4 to 6 feet in length, with a tail from 20–25 feet long. The head appeared doubled round by the tail, and the monster had a huge lion face with large and terrible saucer eyes.[24]

The above "sea devil" sounds quite fanciful, and further complicating the discussions surrounding the serpent's existence was the discovery of creatures never before seen. The president of the New York Lyceum of Natural History announced the capture of a large fish that he described as "truly a monster." It weighed between four and five tons, had two pectoral fins that spanned eighteen feet and from snout to tail was fourteen feet long. In actuality, it was not a fish as it was viviparous and classified as belonging to the family Reja and would perhaps be given its own genus between Squalus and Ascipenser.[25] If this creature was the sea serpent, it would not be evidence of a creature that had survived from the ancient past, and therefore, would be of no use to Lyell in support of his steady-state theory.

While Lyell was visiting North America in the mid-1840s, another rash of sightings occurred both in North America and off the coast of Norway. Visiting Boston in 1845, Lyell saw ads everywhere announcing that a Mr. Koch was exhibiting the fossil skeleton of "that colossal and terrible reptile the sea serpent which when alive measured 30 feet in circumference." It was 114 feet long, with enormous jaws, and its ribs formed an ovoid shaped body with the vestiges of a pair of flippers. Koch had mounted the skeleton in an undulating shape and with its head raised in a position that corresponded to the description from most of the eyewitness accounts of the sightings off the coast of New England. He named the monster *Hydragos sillimannii* in honor of Benjamin Silliman who had given his support in favor of the serpent's existence in 1827. Professor

Silliman was quite familiar with giant fossil reptiles, and argued that the serpent might be a saurian, but that it was not anything like an ichthyosaur. Koch told the public that this fossil was the leviathan of the Book of Job. The fossil had been removed from a bed in Alabama that Lyell later visited. He determined that the fossil was a zeuglodon that Richard Owen had first described as a huge extinct cetacean.[26] One of the people to visit Koch's display was the anatomist Jeffries Wyman who realized it was a hoax, consisting of bones from at least five fossil specimens of the zeuglodon. When Silliman heard that the skeleton was a fake, he demanded that his name not be associated with it. Koch merely renamed it *Hydroachos harlani* after another zoologist, Dr. Richard Harlan.

Koch's fossil may have been a fake, but it was put together from bones of a real creature from the past. More and more fossil zeuglodons were being found along with other strange remains that gave credibility to the sea serpent's existence. Virtually every day, Lyell was being asked if this great fossil skeleton from Alabama was the same as that of the sea serpent that had been seen off the coast near Boston in 1817. At the same time he also received a letter from his geologist friend J. W. Dawson with whom he had explored Nova Scotia in 1841. Dawson was collecting evidence for Lyell of a marine monster that had appeared at Mergomish in the gulf of the St. Lawrence in August 1845. Dawson wrote that the serpent had "been seen by several people very close to the shore." They described it as "being about 200 feet long with an obtuse head like a seal. . . . It was seen to bend its body nearly into a circle, to move in vertical undulations, and to raise its head above the water and maintain it in that position for 15 minutes."[27] Dawson claimed the account was given by respectable, intelligent men, and pointed out that if it was correct the animal must far excel the great reptiles of older times. Similar sightings had occurred in October 1844. Lyell received other letters with comparable descriptions in February 1846. In addition, in July 1845 and August 1846 independent accounts appeared in Norway of a marine animal being seen. Some claimed that it was the same monster that Pontopiddan had seen in 1752.

Lyell had also written a variety of other people including the zoologist G. R. Waterhouse and a Colonel Perkins of Boston who had seen the sea serpent in 1817. Colonel Perkins had written up a detailed account of the serpent and Lyell received a long letter from the man including his notes from the earlier time. Lyell had asked Waterhouse for his opinion of the famous Stronsa beast of 1808 that had been seen off the Orkney Islands. The stronsa beast had caused quite a stir, but eventually most authorities concluded, as did Waterhouse, that it was the remains of a large basking shark.[28] Although Lyell had collected a variety of accounts, he was in an awkward position, confessing "when I left America in 1846, I believed in the sea serpent without having ever seen it."[29]

Of all the sightings of the sea serpent in the nineteenth century, the one that received the most attention was that by Captain Peter M'Quhae of the HMS *Daedalus* in August 1848 off the Cape of Good Hope (see figure 2.2). Accounts of the incident appeared in virtually all the London papers. Lyell had clipped and saved most of them. The captain claimed:

> that had it been a man of my acquaintance I should have easily recognized his features. . . . The diameter of the serpent was about 15 or 16 inches behind the head, which was, without a doubt, that of a snake; . . . its color a dark brown, with yellowish white about the throat. It had no fins, but something like a mane of a horse, or rather a bunch of seaweed, washed about its back. It was seen by the quartermaster, the boatswain's mate, and the man at the wheel, in addition to myself and officers above mentioned.[30]

Lieutenant Edgar Drummond also wrote up his own account that confirmed the captain's description, claiming that the serpent was traveling at a rate of about twelve to fourteen miles an hour. They saw it with the naked eye for about five minutes, but with a glass for another fifteen.

Lyell was not the only scientist who was following the news of the *Daedulus*. The leading British comparative anatomist Richard Owen had a scrapbook that he kept from the 1830s through the 1870s collecting articles and newspaper clippings, all about the sea serpent.[31] But Owen, unlike Lyell, was not kindly disposed to the sea serpent of Captain M'Quhae. The *Annals of Natural History* published the long letter he wrote to the *Times* giving his opinion of the nature of the monster. He pointed out the description of the head was totally unlike what a snake would look like. From all the evidence Owen concluded that what was seen was most likely a large species of seal, probably *Phoca probscidia* or sea elephant. Owen not only discussed the *Daedulus* sighting, but mentioned several other accounts of serpents. One by one, he discredited the idea that they could be relics from the past.

Owen became embroiled in the sea serpent controversy for several reasons. As part of an emerging group of professional scientists, Owen wanted a fundamental redistribution of cultural authority from traditional institutions that included the judiciary, the aristocracy, and the clergy. The sea serpent debates were an important phase in establishing science as the voice of authority in matters of natural history. Owen argued that scientific methodology and expertise were the proper means of establishing the truth. However, the language of the sea serpent literature was usually that of the courts. Judicial procedures including eyewitness accounts, testimonies, and written reports in the form of affidavits and depositions were all means of establishing the legitimacy of the serpent sightings. Owen

FIGURE 2.2. The Great Sea Serpent as First Seen from the HMS *Daedalus*. From *Illustrated London News*, October 28, 1848.

rejected this evidence. Not only did he deliver a devastating critique of M'Quhae's description, but he also dismissed the entire collection sea serpent sightings with the caustic comment that "a larger body of evidence from eyewitnesses might be got together in proof of ghosts than of the sea serpent."[32]

In addition to wanting to remove from the judiciary the authority of deciding what constituted proper evidence, Owen's attack on the sea serpent served another purpose as well. Owen used this controversy to promote his own scientific agenda and discredit Charles Lyell and his antiprogessionist view of earth history. Owen maintained that all the eyewitness accounts were worthless if one did not have corroborating evidence in the form of actual specimens. He asked his readers, "Why have no remains ever been found in spite of extensive searching?" If a sea serpent existed, the species must have been perpetuated through succeeding generations and many individuals must have lived and died. The coast of Norway had been thoroughly explored in search of Krakens and sea serpents, but not one skeleton, not even one bone of a nondescript or indeterminable monster had ever turned up. Norwegian museums were full of skeletons of whales, cachoalots, walruses, and seals, but no monsters. Clearly having Lyell and the antiprogressionists in mind, Owen told his readers that the sea saurians of the secondary period were replaced in the Tertiary by marine mammals. No remains of cetacea had been found in the Lias or Oolite and no remains of a plesiosaur or ichthyosaur or any other secondary reptile had been found in the Eocene or later Tertiary deposits. Since no carcass of these ancient reptiles had ever been discovered in a recent or unfossilized state, it seemed to Owen more probable that men had been deceived by a fleeting view of a partly submerged and rapidly moving animal that might only be strange to themselves.

In his commentary, however, Owen distorted both the captain's comments and Lyell's true position. Captain M'Quhae had never claimed, as Owen had written, that the creature had "a capacious vaulted cranium" or a "stiff inflexible trunk." Owen drew this conclusion since no undulating movements had been observed in the sixty feet of the creature that was seen. Owen did more than try to discredit the eyewitness accounts. His remarks were also meant to demonstrate the implausibility of Lyell's view of earth history.

Lyell was interested in the sea serpent because it would have provided evidence for his theory of climate, which in turn provided an explanation for his antiprogressionist view of the fossil record. Lyell claimed that local climate was determined by the relationship of the land to the sea, on the position and size of mountain ranges, and on winds and ocean currents. Due to geological processes such as erosion and earthquakes, the distribution of land and the sea changed, which in turn led to changes in the climate. His research indicated that geological and geographic changes did not seem directional at all. Rather, they appeared to be cyclical just as Hutton had proposed. Since species were adapted to their environmental conditions that were cyclic and recurrent, how could there be progression? As he enthusiastically wrote to a colleague his "grand new theory would . . . give you a receipt for growing tree ferns at the pole, or if it suits me pines at the equator; walruses under the line and crocodiles in the Arctic Circle." Moreover, he continued, "All these changes are to happen in the future again, and iguanodons and their congeners must as assuredly live again in the latitude of Cuckfield as they have done so."[33]

Ironically, while Lyell's theory of climate was what made him interested in sea monsters as ancient relics from the past, the theory also led ultimately to his dismissing their survival to the present as a real possibility. If organisms were so closely tied to their environmental conditions, of which climate was the most important, then organic change depended directly on climatic change. Lyell agreed that the fossil evidence indicated the Carboniferous period had a tropical climate. This was problematic for his steady-state theory and his solution was quite radical. He did not deny the present was cooler, but he emphasized that the geological record represented only a small part of a complete cycle. This still would leave intact the idea that the iguanodon might someday return. Lyell certainly recognized that environmental conditions were so complex that they would never repeat exactly. Thus, he did not think the exact species of iguanodon would return, but only the genus. Nevertheless, if the climate had truly changed, even if the change was cyclic, it would mean that organisms would not survive from a period of earth history when the climate was tropical. While Lyell may have believed that someday the plesiosaur might return, he also thought it was quite unlikely that a present-day sea serpent would be a reptile. In the northern latitudes where the serpent was most often sighted, reptiles were quite rare. No large reptiles existed in the immediate geological period, but in the colder latitude, like today, many huge sharks, narwhals, and whales could be found. He thought that a creature might exist that was some unknown species of one of these above mentioned families and he saw "no impropriety in its retaining the English name sea serpent."[34] Therefore, there was no reason to think a sea serpent could not exist, but he did not think it would be a reptile. Once again, Lyell used the same argument as when he was trying to make his case against progression: not much weight should be given to negative evidence. He pointed out that zoologists had noted that some whales were so rare they hadn't been seen since the seventeenth century and only three specimens had been found of another species.

Nevertheless, Lyell's claim that we had a fossil record of only one phase of a great geological cycle was highly speculative. He had to make an inference about future conditions that by necessity was based on no evidence at all—and this represented the kind of theorizing that he claimed to be so against.[35] The existence of a sea serpent would extricate him from this quandary. Thus, even when he later concluded that the sightings along the North Atlantic coast were probably not of a sea serpent, but most likely of *Squalus maximus* or the basking shark, he continued to follow reports of other sightings. If an organism were found, even if it was a different species, but still belonged to a genus that was supposedly extinct, it would be a great boon to his antiprogressionist position, which he argued for throughout the 1850s.

Although this account makes Lyell's interest in the sea serpent reasonable, it is incomplete. Earlier, I claimed that Lyell became the serpent's most eminent advocate, but that is not quite accurate. Although Lyell had an entire separate file devoted to sea serpents in his scientific papers, he never referred to them in his extensive scientific correspondence or in any published material (with the exception of his *A Second Visit to the United States of North America*). Lyell marginalized his own interest in sea monsters. Why was he so reluctant to link the possible existence of sea monsters to his theory of climate and the history of life? The answer lies in situating sea monsters within the developments occurring in the newly created disciplines of geology and paleontology.

Sea Serpents and the Professionalization of Geology

Given the recent finds in paleontology it was not surprising that Victorians became enamored with the sea serpent. As Cuvier wrote of the newly discovered fossil reptiles: their combinations of structure "astonished naturalists [and] without the slightest doubt would seem incredible to anyone who had not been able to observe them himself. The *Plesiosaurus* is perhaps the strangest of the inhabitants of the ancient world and the one which seems most to deserve the name of monster."[36] Sea serpents now had the potential of being integrated into a totally naturalistic account of the history of life. A dramatic increase in serpent sightings occurred that correlated with the discovery of fossil remains. Before 1639 there was only one dated report, then four between 1650 and 1700, and four more between 1700 and 1750. In the last half of the eighteenth century there were twenty-three sightings, but from 1800 to 1850 there were 166 documented sightings. The latter part of the nineteenth century saw only a slight decrease to 149.[37] One could read about the serpent in a hobbyist magazine such as *Field and Stream*, in general periodicals such as *Blackwood's*, *Edinburgh Magazine*, and the *Westminster Review*, as well as in prestigious journals such as *Nature* and *Zoologist* (see table 2.1). Impassioned debates took place as to what kind of organisms these sightings represented and what they could be mistaken for.

Table 2.1. Nineteenth-Century Journal Articles on Sea Serpents

1808	Barclay, J. "Remarks on Some Parts of the Animal that Cast Ashore on the Island of Stronsa, Sept 1808." *Memoirs, Wernerian Natural Hist. Soc.* 1: 418–44.
1817	Linnean Soc. of New Eng. *Report of a Committee of the Linnean Soc. of New eng. Relative to a Large Marine Animal Supposed to Be a Serpent. Seen near Cape Ann, Mass in Aug. 1979.* boston: Cumming & Billiard.
1819	"American Sea Serpent." *Philosophical Magazine* 53: 71.
1820	Bigelow, J. "Documents and Remarks Respecting the Sea Serpent." *American J. of Sci.* 2: 147–64.
1821	"Analysis of One of the Vertebrae of the Orkney Animal." *Edinburgh Phil. J.* 5: 227.
1823	Mitchill, Samuel. "The History of Sea-Serpentism." *American J. Sci.* 15: 351–56.
1827	Hooker, W. J. "Additional Testimony Regarding the Sea Serpent of the American Seas." *Edinburgh J. of Sci.* 6: 126–33.
1834	"Saurian? Reptiles: Recent Appearance of the Great American Sea Serpent." *Magazine of Natural History* 7: 246.
1845	Hamilton, R. *Amphibious Carnivora.* In *The Naturalists Library: Mammalia.* W. Jardine. Edinburgh: W. H. Lizars vol. 25, pp. 313–26.
1845	Koch, Albrecht K. *Description of the* Hydrachose harlani: *Koch* 2d ed. New York: B. Owen, 24 pp.
1845	Wyman, Jeffries. "The Fossil Skeleton Recently Exhibited in New York as That of a Sea-Serpent under the Name of *Hydrarchos Sillimani.*" *Proc. Boston Soc. of Nat. Hist.* 2: 65–68.
1847	"A Plea for the North Atlantic Sea-Serpent." *Zoologist* 5: 1841–46.
1847	Deinboll, P. W. "The Great Sea Serpent." *Zoologist* 5: 1604–08.
1847	Newman, Edward. "Preface." *Zoologist* 5: x–xi; 6: xi–xii.
1847	Sullivan, W., et al. "The Sea Serpent." *Zoologist* 5: 1714–15.
1848	Cogswell, Charles. "The Great Sea Serpent." *Zoologist* 6: 2316–23.
1848	Cooper, W. W. "The Great Sea-Serpent." *Zoologist* 6: 2192–93.
1848	Drummond, Edgar. "The Great Sea-Serpent." *Zoologist* 6: 2306–07.
1848	Meville, A. G. "The Great Sea-Serpent." *Zoologist* 6: 2310.
1848	Woodward, J. "The Great Sea-Serpent." *Zoologist* 6: 2028.
1848	"The Great Sea-Serpent." *Zoologist* 6: 2307–16, 2323–24; 7 (1849): 2458–60; 21 (1863): 8727.
1848	"The Great Sea-Serpent." *Scientific American* 4: 66.
1848	Owen, Richard. "The Great Sea-Serpent." *Annals and Magazine of Nat. Hist.* ser. 2, 2 458: 63.

(Continued)

TABLE 2.1. (*Continued*)

1848	Perkins, T. H. "The Great Sea-Serpent." *Zoologist* 6: 2359–63.
1849	Davidson, R. "The Great Sea-Serpent." *Zoologist* 7: 2458–59.
1849	"A Strange Marine Mammal." *Zoologist* 7: 2433–34.
1849	Lyell, C. *A Second Visit to the United States of North America* 2 vols. London: John Murray, vol. 1: 132–40.
1849	Newman, Edward. "Enormous Undescribed Animal, Apparently Allied to the Enalisauri, Seen in the Gulf of California." *Zoologist* 7: 2356.
1849	Newman, Edward. Probability of the Present Existence of Enaliosaurians." *Zoologist* 7: 2395.
1849	Newman, Edward. "The Great Sea-Serpent An Essay, Showing Its History, Authentic, Fictitious, and Hypothetical." *Westminster Review* 50: 491–515.
1849	Newman, Edward. "Inquiries Respecting the Bones of A Large Marine Animal Cast Ashore on the Island of Stronsa in 1808." *Zoologist* 7: 2358–59.
1849	*An Essay on the Credibility of the Existence of the Kraken, Sea Serpent and Other Sea Monsters*. London: W. Tegg. 61 pp.
1849	"A Young Sea-Serpent." *Zoologist* 7: 2395–96.
1850	Newman, Edward. "The Great Sea-Serpent." *Zoologist* 8: 2925–58.
1850	Newman, Edward. "The Great Sea-Serpent." *Zoologist* 8: 2803–24.
1852	"Reported Capture of the Sea-Serpent." *Zoologist* 10: 3426–29.
1853	Newton, Alfred. "The Sea Serpent." *Zoologist* 11: 3756.
1853	Steele, T. "The Great Sea-Serpent." *Zoologist* 11: 3756.
1854	Traill, T. S. "On the Supposed Sea-Snake, Cast on Shore in the Orkneys in 1808 and the Animal Seen from the HMS *Daedalus* in 1848. *Proc. Roc. Soc. of Edinburgh* 3: 208–15.
1856	Baird, Spencer. "The Sea Snake Story a Fiction." *Zoologist* 14: 4998.
1856	More, A. G. "The Sea Serpent." *Zoologist* 14: 4968.
1857	Baillie, J. "Jewish Tradition Respecting the Sea Serpent." *Notes & Queries* ser. 2, 3: 149.
1857	Eastwood, J. "Jewish Tradition Respecting the Sea Serpent." *Notes & Queries* ser. 2, 3: 336.
1858	Fraser, William. "Jewish Tradition Respecting the Sea Serpent." *Notes & Queries* ser. 2, 6: 277–78.
1858	van Lennep. "A Nine Days Fight with a Sea-Monster." *Notes & Queries* ser. 2, 6: 524.
1858	Harrington, G. H. "Another Peep at the Sea Serpent." *Zoologist* 16: 5989.
1858	Smith, Alfred Charles. "The Sea Serpent." *Zoologist* 16: 6015–18.
1858	Smith, Fredrick. "The Sea Serpent." *Zoologist* 16: 5990.
1859	van Lennep, J. H. "Another Sea Serpent." *Zoologist* 17: 6492.

TABLE 2.1. (*Continued*)

1860	Adams, A. "On the Probable Origin of Some Sea Serpents." *Zoologist* 18: 7237.
1860	Newman, Edward. "Captain Taylor's Sea Serpent." *Zoologist* 18: 7278.
1860	Newman, Edward. "Note on an Ophioid Fish Lately Taken in the Island of Bermuda." *Zoologist* 18: 6989–93.
1860	Taylor, W. "The Sea Serpent." *Zoologist* 18: 6985–86.
1862	Cave, Stephen. "The Sea Serpent." *Zoologist* 20: 7850–51.
1863	Baring-Gould, S. *Iceland: Its Scenes and Sagas*. London: Smith, Elder. pp. 345–48.
1867	Buckland, F. *Curiosities of Natural History, Second Series*. London: Richard Bentley. pp. 217–20.
1870	"The Sea Snake Again." *Land and Water* Apr. 9.
1872	"Notes." *Nature* 6: 130, 270, 402; 21 (1881): 565; 32 (1885): 462.
1872	"Remarks on T. T. Letter." *Land and Water* Sept. 7, p. 158.
1873	Francis, F. and Joass, James. "The Sea Serpent." *The Field* 42: 511 Nov. 15.
1873	Macrae, J. and Twopeny, D. "Appearance of an Animal Believed to Be That Which Is Called the Norwegian Sea Serpent on the Western Coast of Scotland, in August, 1872." *Zoologist*, ser. 2, 8: 3517–22.
1873	Newman, Edward. "The Supposed Sea-Serpent." *Zoologist*, ser. 2, 8: 3804.
1875	Kilgour, H. "The Great Sea Serpent." *Notes & Queries* ser. 5, 4: 485–86.
1876	Dyer, High McN. "The Mythical Sea-Serpent?" *Science Gossip* 12. 260.
1877	"Occurrence of a Sea Serpent." *Land and Water* Sept. 8 & 196–98.
1877	Cornish, T. "The Great Sea-Serpent." *Land and Water* Sept. 15.
1877	Nahanik. "The Great Sea-Serpent." *Land and Water* Sept. 15, p. 220.
1877	MacDonald, Daniel. "The Sea-Serpent." *Land and Water* Sept. 15, p. 263.
1877	Wilson, A. "The Sea-Serpents of Science." *Land and Water* Sept. 15, pp. 218–20
1877	"The Great Sea-Sepent." *Land and Water* Oct. 27, p. 355.
1878	"Supposed Sea Snake Caught in Australia." *Land and Water* May 24.
1878	Bird, C. "The Sea-Serpent Explained." *Nature* 18: 519.
1878	Drake, Joseph. "The Sea-Serpent Explained." *Nature* 18: 489.
1878	Ingleby, C. M. "The Sea-Serpent Explained." *Nature* 18: 541.
1879	Barnett, H. C. "The Sea-Serpent Explained." *Nature* 20: 289–90.
1880	Wood, Searles V. "Order Zeuglodontia, Owen." *Nature* 23: 54–55, 338–39.
1883	Aldos, W. S. "The Sea Serpent." *Nature* 23: 338.
1883	Barefoot, William. "The Sea Serpent." *Nature* 27: 338.
1883	Gibson, J. *Monsters of the Sea*. London: T. Nelson.

(*Continued*)

TABLE 2.1. (*Continued*)

1883	Mott, F. T. "The Sea Serpent." *Nature* 27: 293–94.
1883	Sidebotham, J. "The Sea Serpent." *Nature* 27: 315.
1884	Lee, H. *Sea Monsters Unmasked.* London: William Clowes, pp. 52–1–3.
1886	Colonna, B. A. "The Sea Serpent." *Science* 8: 258.
1886	"Contributions to a Bibliography of the 'Sea Serpent'." *Proc. Roc. Phys. Soc. Edinburgh* 9: 202–205.
1887	Stejneger, Leonard. "How the Great Northern Sea-Cow Rytina Became Exterminated." *Amer. Naturalist* 21: 1047–54.
1890	Ashton, J. *Curious Creatures in Zoology.* London: J. C. Nimo pp. 226–35, 245–49, 268–78.
1892	"Another Sea Serpent." *Scientific American Supplement* 34: 14115.
1895	"Another Sea Serpent." *Scientific American* 73: 21.
1897	Lucas, F. "The Florida Monster." *Science* 5: 467.
1898	Ballou, W. "Serpent-like Sea Saurians." *Popular Science* 53: 209–25.

Source: Adapted by author from bibliographic information from G. M. Eberhart, Monsters (New York: Garland, 1983).

There was no shortage of explanations that included sea snakes, rows of porpoises or seals, giant eels, basking sharks, whales, seaweed, floating dead trees, and flights of low lying birds. Many speculated that the sightings were possibly a *plesiosaurus*, an *ichthyosaurus*, or an ancestral whale, such as the *Basilosaurus*. The serpent would no longer be the stuff of legend, but a real creature, that had survived from the past. Nevertheless, sea serpents remained at the fringes of scientific respectability, as they became the locus for two different but related boundary disputes that were occurring in the struggle for geology to achieve professional status.

First, sea serpents were so closely tied to mythology that many people considered investigation of them as pseudoscience. Even if mermaids had been dismissed from serious discussions about inhabitants of the sea, this did not prevent the public, both scientists and nonscientists, to fill the void with other mythological creatures. One has to make a distinction, however, between those people who believed in the sea monster of ancient legend and myth, and those like Lyell who thought that once found it would be similar to known animals, that is, it might be a fish, a reptile, or mammal and it would look similar to a living or, as Lyell hoped an extinct organism. However, often this distinction was blurred, casting a shadow on the serpent's status as a serious object worthy of study.

Second, geology had been a subject practiced primarily by amateurs. Joseph Anning found the first ichthyosaur in 1811 in Lyme Regis and his sister Mary found the remainder of the skeleton twelve months later. Mary

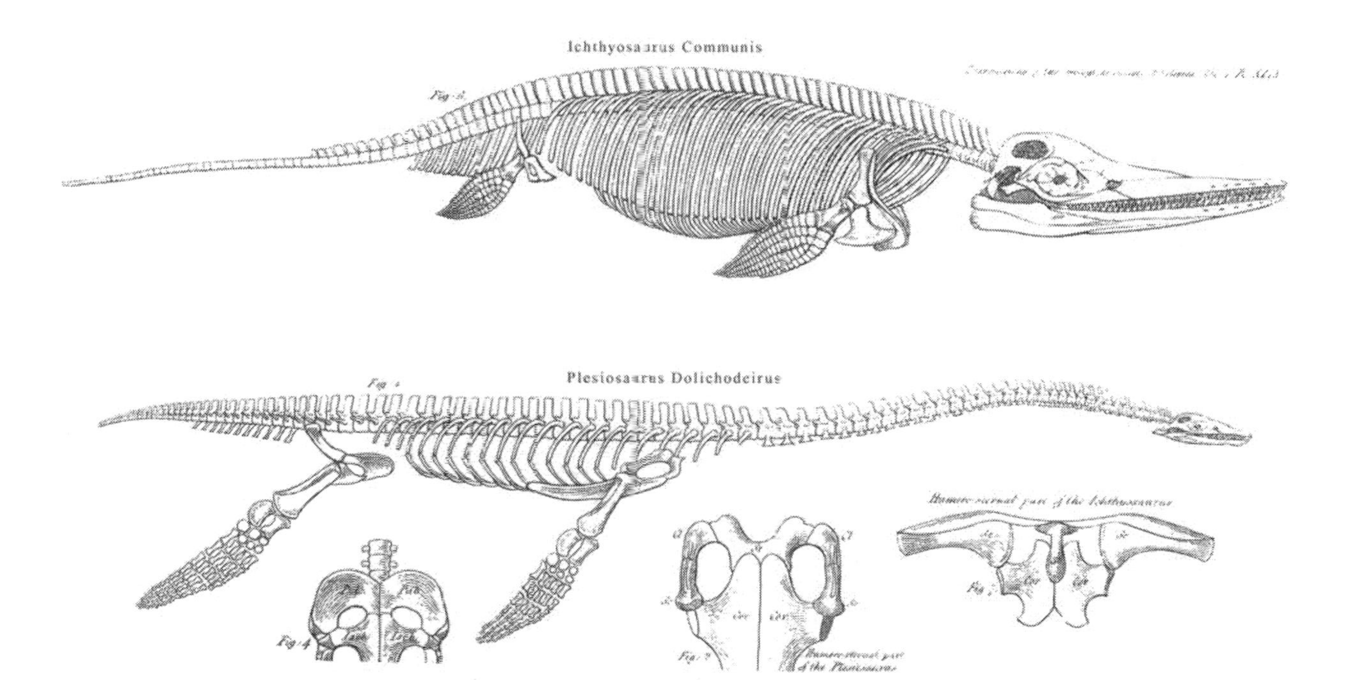

FIGURE 2.3. William Conybeare's Reconstructions of the *Ichthyosaurus* and *Plesiosaurus* Based on Skeletons Found in the Liassic Strata. From William Conybeare, "On the Discovery of an Almost Perfect Skeleton of *Plesiosaurus*," *Transactions of the Geological Society of London*, 2nd series, 1824. Published in M. J. S. Rudwick, *Scenes from Deep Time: Early Pictorial Representations of the Prehistoric World* (Chicago: University of Chicago Press, 1992). Reprinted by permission of the publisher.

also discovered the first complete plesiosaur and Britain's first pterodactyl. Mary was the most famous fossil hunter of her time and in 1838 she was given an annual stipend from the British Association for the Advancement of Science (BAAS). Nevertheless, she was the exception and fossil hunters for the most part only earned money by selling what they had found. If geology was to be taken seriously and regarded as a profession worthy of paid positions, a clear line had to be drawn between who was a professional and who was not. People who had no training in scientific methodology made the vast majority of serpent sightings. Credibility was based on the 'unimpeachable character' of 'witnesses' who made depositions and signed affidavits testifying to the truth of their observations. Labeling these people as amateurs was one way to discredit the validity of their observations and redefine what counted as proper evidence. Sightings alone would not be enough.

Serpent sightings were beset with numerous cases of mistaken identity, even if these did not represent deliberate hoaxes. A particularly embarrassing case involved the American Linnaean Society. At the time Lyell was in Boston, a snake about three feet long and having a series of humps washed ashore and was brought to the attention of the Linnaean Society. Many claimed it was an exact miniature of the sea serpent of which there had been numerous sightings in 1817, and which had recently been seen off the coast of Boston. The Linnaean Society prepared a detailed monograph on the specimen, declared it must be the young of the famous 1817 sea serpent and gave it the impressive name of *Scoliophys atlanticus*. The identification of the serpent helped legitimize the society as a scientific organization and after it disbanded, many of its members went on to establish the Boston Society of Natural History.[38] But much to the society's dismay, Louis Agassiz later identified the specimen as a full-grown common New England species of land snake—the humps on its back due to a diseased spine. Undoubtedly, incidents like this contributed to the difficulty in establishing the credibility of the sea serpent. Nevertheless, the practice of science is full of misconceptions and mistakes. Mistakes are not enough to irrevocably harm a particular area of investigation. In spite of being responsible for discrediting the Linnaean specimen, Agassiz still believed in the serpent's existence. Charles Cogswell, a well-respected amateur naturalist, in his "Plea for the North-Atlantic Sea Serpent" acknowledged that there were many hoaxes and false claims, but lamented that so many people dismissed the evidence.[39] How did these descriptions compare with what the ancient saurians were believed to look like from the fossil remains? How could these organisms survive for such a long period of time, especially if climatic conditions have changed? Articles tied the search for the serpent with other types of investigations of the day. As the anonymous author of the 1849 article "An Essay of the Credibility of the Existence of the Kraken, Sea

Serpents, and other Sea Monsters" stated, "There are no monsters now a days . . . and incredible as the vague accounts of unclassified forms present themselves . . . we may rest assured that if we could bring them within the reach of scientific examination, the supposed monsters would readily arrange themselves in the formulae of systematic arrangement."[40] He believed that the serpents would fill the gap between the cartilaginous fishes and amphibious reptiles—in particular the plesiosaurs. Agassiz and Lyell would have been sympathetic to such a position. However, vindication of this claim depended on finding an actual specimen.

In 1852, the *Zoologist* published "Reported Capture of the Sea-serpent," an article by Captain Charles Seabury of the American whaling ship *Monongahela*. The crew had sighted a serpent while sailing the South Pacific "moving like the waving of a rope when held and shaken in the hand." Determined to catch it, the captain wrote, "Our courage was at stake—our manhood, and even the credit of the American whale-fishery." They chased the creature and in spite of having several harpoons stuck in it, the serpent towed the boat for sixteen hours before they were able to actually capture it. They dried the 103 foot long blubber covered snake, preserved its head, which was described as "long and flat with ridges," and having 94 teeth, "in the jaws, very sharp, all pointing backward and as large as one's thumb."[41] They continued sailing, promising to finally deliver tangible proof of the sea serpent's existence. Reports of this encounter were passed along to other ships and then finally printed in the *New York Tribune* and copied by the *Times*. The *Zoologist* printed the densely packed four-page article, followed by an explanation of its origin. However, the journal's editor commented that he thought it was probably a hoax, although quite well done. The ship was lost in the Arctic in 1853 with no survivors, and thus, no one was ever able to verify that a serpent had indeed been captured. Once again, the sea serpent was not able to swim out of the ocean of tall tales into the sea of reality.

While the discovery of fossils brought a renewed interest in the sea serpent, at the same time they essentially guaranteed that the serpent would not be a part of geological debates that were occurring all over the country. Whether it was at the Geological Society of London or its local equivalents in the provinces, or universities such as Oxford and Cambridge, or the BAAS, fossils dominated the discussions. The practices developed to unearth, preserve, transport and display fossils were key to the professionalization of geology in the nineteenth century. Collections of fossils played a critical role in the establishment of museums as scientific institutions.[42] Unlike the intangible serpent, they were real objects that could be collected, measured, compared, and needed interpretation.

With several emerging scientific disciplines vying for power and authority, the elusiveness of the serpent meant it would remain at the fringes of scientific investigation. The term scientist had only recently been introduced by William Whewell and represented a group of people who thought they should have a greater say in the running of society. Often the boundary between amateur and professional was not sharp. While the gentlemen of science may not have earned their living as scientists, they had a definite view of science that they wanted to promote. In the early years of the BAAS, aristocratic approval was essential for its success. Not only were aristocrats encouraged to become members of the BAAS, but many of its founding members thought the association could best further its goals by having a local aristocrat to preside as president of the annual meetings. However, many of the leading gentlemen of science deplored the great adulation given to the dukes and other titled gentry. Lyell believed that a man of science for the 1841 presidency "is the only thing that can redeem our proceedings from the reproach of taking a very low standard in the estimate of the real dignity of scientific men."[43] Lyell's view prevailed and independent gentlemen who espoused a particular ideology of science overshadowed both aristocrats and clergy. They wanted distinct boundaries drawn between natural, and religious or political knowledge. Science was to be clearly demarcated from other kinds of knowledge and represented a value neutral objective means of inquiry.[44]

Whewell went even further in defining scientific practice. It was not just the collection of facts, but rather facts combined with theory. George Harcourt, another prominent member of the BAAS, had suggested that local observers, even experimenters with modest ability and talent could make valuable contributions to science by collecting and recording data. Whewell disagreed, arguing, "A combination of theory with facts, of general views with experimental industry is requisite, even in subordinate contributions to science."[45] Once facts were classified and connected to theory, they became knowledge, and if they became embedded as part of some general propositions they then became truth. Thus, Whewell denied that provincial meteorologists would help the BAAS just by recording data, claiming "we might notice daily changes of the winds and skies, . . . which shall have no more value than a journal of dreams."[46] If Whewell had no use for people accumulating meteorological data, there was certainly no place for amateur naturalists gathering information on sea serpents whose very existence was questionable.

With various kinds of inquiry jockeying for recognition as "science," geologists in particular wanted to remove their investigations of natural history from that of myth and legend. Yet it certainly is not surprising that comparisons would be made between the fantastic fossil remains that were being discovered almost daily and the mythological creatures of ancient legends. Perhaps those creatures did have some basis in reality. Anthropologists were debating the meaning of myth and legend, some arguing that they represented an actual, if embellished event in the past and should be regarded as history. It might seem reasonable for paleontologists to look for evidence of these "mythical" creatures in the form of present-day sea monsters, but paleontology was a new discipline, struggling for recognition as a legitimate science. It was also a "gentleman's science." Most members of the Geological Society of London were men of independent means and able to study geology for its own sake. Few people made their living as geologists or paleontologists, and thus, defining who was a professional was problematic. Conybeare was a country parson living in the area where the first plesiosaurs and ichthyosaurs were found. He had an avid interest in natural history that had begun when he was a young Oxford don in the early days of the Geological Society. Like most paleontologists, he adopted Cuvier's technique of using comparative anatomy to make his fossil reconstructions. His *Outlines of Geology of England and Wales* (1824) quickly became the authoritative reference on stratigraphical succession and he became well known to geologists both home and abroad. Conybeare was typical of the gentleman scientist who was setting the agenda for paleontology of what would be acceptable topics and who would be regarded as an authority.

One way that Owen tried to discredit the eyewitness accounts of sea serpents was by asking whom should we believe, a mariner or a professor of comparative anatomy? But that *was* precisely what people contested. Edward Newman, a highly respected amateur naturalist, claimed, "the majority of our professors and curators would not know a whale from a porpoise, a shark from an *ichthyosaurus* if they beheld these creatures in their native element." The professor's expertise was only relevant when creatures were stuffed with straw or reduced to skeletons or a few fragments were examined under a microscope.[47] Newman would much rather trust the observations of the whaler, mariner, or harpooner—the practical fisherman. It was an insult to people who made their living off the sea to claim that they wouldn't know the difference between a school of seals or porpoises, a flock of birds, and a sea serpent.

Newman recognized that a shaky consensus had emerged in the professional geological community and an unequivocal identification of a genuine sea serpent—that is a relic from the past—would upset the prevailing hypothesis that the history of life was progressive. This in turn would necessitate numerous modifications of accepted scientific dogma. Lyell's steady-state views also challenged current scientific orthodoxy. The serpent would have provided outstanding evidence for his antiprogressionist views, but his minority position had already been the subject of ridicule.

In 1831 Lyell was appointed the first professor of geology at the new Kings College in London, but most geologists found his cyclic view of earth history totally implausible. One passage in the *Principles* seemed especially outrageous:

> There would be a preponderance of tree-ferns and plants allied to palms and arborescent grasses in the isles of the wide ocean. . . . Then might those general of animals return, of which the memorials are preserved in the ancient rocks of our continents. The huge iguanodon might reappear in the woods, and the ichthyosaur in the sea, while the pterodactyle might flit again through umbrageous groves of tree ferns.[48]

Henry De la Beche, who had coauthored Conybeare's description of the plesiosaur, drew a cartoon entitled "Awful Changes," created as an in-joke for the members of the Geological Society of London (see figure 2.4). In it Lyell was portrayed as Professor Ichthyosaurus lecturing on a human fossil skull before an attentive audience of ichthyosaurs and plesiosaurs.[49]

Lyell did not want his theories subjected to further criticism by having them associated with mythological creatures and a public who was all too quick to embrace the fantastic. He was unwilling to go public with his views and even remained guarded in his private conversations with fellow geologists concerning his interest in the sea serpent. Owen, however, eagerly entered the fray about sea monsters because it allowed him to promote his view of earth history that represented those of the scientific establishment. Equally important, the sea serpent controversy provided a forum to extol the virtues of the scientific method.

According to judiciary standards, the numerous eyewitness accounts provided compelling evidence that demanded sea serpents be taken seriously. For Owen it didn't matter how many affidavits and depositions had been taken in support of the serpent. Scientific practice demanded that the sightings had to be matched with corroborating specimens in the collections. New unfamiliar organisms had to be analyzed in terms of existing specimens. Not only were there no specimens to analyze, but also the idea of an ancient reptile surviving from the past was counter to the fossil record concerning the history of life. There was no evidence of those ancient saurians surviving even into the next geologic epoch, much less into the present. Owen brought home this point in his contribution "Zoology" for the Admiralty's *Manual on Scientific Inquiry* of 1849. In documenting phenomena that could be taken as evidence of the sea serpent's existence he recommended that whenever a large strange creature appeared in the ocean that it should be referred to as an "animal" rather than a "sea-serpent." He also provided specific instructions for making observations and collecting that served to make science rather than the judiciary the arbitrators in evaluating the evidence.[50]

Owen's dismissal of the eyewitness accounts cannot simply be regarded as an interdisciplinary squabble between the emerging scientific community and the legal profession over who should have professional

FIGURE 2.4. Henry De la Beche's Cartoon "Awful Changes" (1830) (lithographed broadsheet). Charles Lyell portrayed as Professor Ichthyosaurus in a parody of his antiprogressionist views concerning the fossil record. Published in M. J. S. Rudwick, *Scenes from Deep Time: Early Pictorial Representations of the Prehistoric World* (Chicago: University of Chicago Press, 1992). Reprinted by permission of the publisher.

authority over matters of natural history. Ultimately, the kinds of objections that Owen raised, particularly the lack of actual specimens undermined the credibility of eyewitness accounts. The legal system also recognized that eyewitness accounts were notoriously unreliable. Five different witnesses may often give five different accounts of what they observed with no intention to deceive. Sea serpent sightings were plagued not only by this problem of perceptual distortion, but also by what is referred to as contagion. Once a report of a sighting appeared in a newspaper or by word of mouth inevitably many others followed. Believers in the serpent's existence claimed that was because a creature did indeed exist, and therefore many people observed it. In the period between 1817 and 1819 several hundred people claimed to have observed the serpent off the north Atlantic coast. James Prince, a district marshal stated that he and about two hundred witnesses watched a sea serpent off Nahant, Massachusetts, for more than three hours in 1819.[51] It is hard to believe that so many people just experienced a collective delusion. However, after the *Daedalus* sighting was reported, numerous others appeared and at least one was an outright fraud, written by a fictitious captain of a ship that did not exist.[52] Because of such chicanery many people worried that their claims would be disbelieved. Sailors often kept their sightings to themselves for fear of ridicule.[53] Thus, supporters of the serpent argued that if anything, serpent sightings were actually under reported. Further complicating the issue was the role that nationalism played in evaluating the reliability of the eyewitness accounts. The London press took seriously the sightings of British naval officers, but dismissed the American sightings off Cape Ann as a Yankee trick.[54] For the British, English eyewitness accounts were usually regarded as more reliable because they were most often made by officers of the crown whereas the Americans were known for "gross exaggerations" or "hoaxing inventions." Philip Henry Gosse, another amateur naturalist, believed in the reality of the serpent, but he also emphasized great care had to be taken in obtaining evidence.[55]

With deliberate hoaxes, the problem of perceptual error, and many sightings that were treated as genuine, but were probably fraudulent, investigations of the serpent did not seem a very promising line of scientific investigation. While knowledgeable and well-respected people, many who would describe themselves as naturalists, wrote many of the articles on sea monsters; nevertheless, they would not be considered professional scientists. Even among those who were regarded as professional, relatively few earned their living as scientists. Lyell was independently wealthy and free to follow his curiosity wherever it led him. For struggling paleontologists who were trying to establish a career that actually would pay, investigations in fringe areas were a risky path to follow. The sea serpent had too many liabilities to be fully embraced by individuals who wanted clear boundaries drawn between professional and amateur and by the emerging disciplines of paleontology and geology whose very subject matter was in many ways marginal.

Not only were physical theories of earth history highly speculative, but also the study of fossils was particularly problematic. It was not until the end of the nineteenth century that fossils were defined solely as remains of once living organisms. The original meaning was simply "dug up" and fossils were described in the context of mineral ores, natural crystals, and useful rocks. The change in the definition of fossil was not trivial as it provides a clue to the first major problem in the history of paleontology. It was not just determining whether fossils were actually organic in origin or not, or recognizing that they had an "obvious" resemblance to living animals and plants, and then deciding what was the 'correct' and what was the 'erroneous' opinion. Rather, it was a much more subtle debate about the meaning and classification of the whole spectrum of 'fossil' objects.[56] It was difficult enough to try to make sense of the wealth of fossils and stratigraphical data without further complicating the view of earth history by searching for sea serpents, creatures that in the final analysis might still properly belong to the realm of imagination rather than the domain of the scientist.

The story of the sea serpent illustrates that facts and their interpretation are continually being negotiated. The discovery of fossil "monsters" played a key role in that negotiation. By the end of the eighteenth century, naturalists were convinced that human history represented only a minuscule part of earth history. Fossils provided a window in which to view this period of prehuman history, making it both possible and legitimate for geologists to create scenarios about the deep past. Most of these fossils clearly had no counterpart to present-day organisms, and resulted in a major reinterpretation of the history of life. Within that revised account, the sea serpent as a real organism that had survived from the past resurfaced from the depths of sea lore as a potential link between myth and science.

Nevertheless, sea serpents did not become a part of mainstream scientific investigations. Their fate is an example of how the increasing power and prestige of science in Victorian England resulted in it gaining a virtual monopoly over the proper means of establishing the "truth." Science has been so successful, in part, because in the words of Peter Medawar it has dealt with, "the art of the solvable." It can never be proven absolutely that something does not exist. However, the lack of specimens combined with what was known about the history of life from the fossil record made the idea of an ancient creature surviving into the present extremely unlikely. This does not mean that every sighting was accounted for, but it does mean that the elusive serpent did not fall into the category of solvable problems. Although scientific developments were in large part responsible for the sea serpent's popularity and legitimacy, paradoxically those developments led ultimately to its demise as a topic of serious investigation. Yet, as late as 1893, even the arch empiricist Thomas Huxley wrote to the *Times*, "there is no *a priori* reason that I know of why snake bodied reptiles from 50 feet long and upward should not disport themselves in our seas as did those of the Cretaceous epoch which geologically speaking is a mere yesterday."[57] George B. Goode and Tarleton H. Bean wrote in *Oceanic Ichthyology*, (1895) the bible of American marine biology at the time, "It cannot be doubted . . . that somewhere in the sea at an unknown distance below the surface, there are living certain fishlike animals, unknown to science and of great size, which come occasionally to the surface and give a foundation to such stories as those of the sea serpent."[58]

The serpent refused to be relegated to the realm of myth, continuing to exert its influence on oceanographic studies. During the Dana Expedition (1928–1930) a giant larva was found that was thought to belong to the eel genus. If the ratio of the larva to the adult were similar to that of the common eel then the adult would be over thirty meters long and provide a factual basis to the myth of the sea serpent.[59] Anton Brunn, who led the Dana expedition, wrote in 1941, "I believe in the sea serpent" claiming there were "probably eel-like sea monsters living at very great depths."[60] He gave public lectures on the possible existence of the sea serpent. Throughout the 1940s articles appeared in Danish newspapers and various members of the scientific community debated the merits of an expedition in search of the serpent. After the war, research money was scarce and oceanographers were trying to gain an appropriation to purchase a vessel and convert it for a two-year project. According to Danish oceanographer Torben Wolff, the many newspaper articles that speculated on the adult form of the Dana larva played a role in persuading politicians to grant funding for the *Galathea* Deep-Sea Expedition.[61] It was not until 1970 that the Dana larva was unequivocally identified, as belonging to another fish order Notacanthiformes, in which the larvae is much closer to the adult form in size, and in some species even larger, thus putting to rest the idea that it could be the larva of the sea serpent. Just as in the nineteenth century, the sea serpent continues to capture the imaginations of not just a gullible public, but leading scientists as well. As scientists continue to explore the universe the border between fact and fantasy continually shifts.

Victorians and the Lure of the Sea Serpent

Given this blurring of boundaries between science and fantasy, it is not surprising that the best contemporaneous account of the sea monster's status in the Victorian period was a work of science fiction. It also suggests an additional reason why Victorians were so invested in proving the serpent's existence.

> The year 1866 was marked by a strange incident, an unexplained and inexplicable phenomenon . . . which disturbed the maritime population and exciting the public mind in the interior of the continent. . . . For some time a number of ships had encountered an enormous thing, a long object spindle-shaped, at times phosphorescent, and infinitely larger and more rapid in its movements than a whale.[62]

Thus begins Jules Verne classic tale of adventure, *Twenty Thousand Leagues Under the Sea*. Published in 1870, the novel began with reports of an unknown monster sinking a number of ships. Verne's description of the sightings by various ships, the rumors, the debates over what kind of creature it was, and its very existence accurately reflects the sea serpent periodical literature of the time.

Verne drew on the Victorian desire to bring phenomena that were regarded as supernatural under the purview of scientific explanation. The novel was sprinkled with passages that could pass as an oceanographic text; but more importantly, Verne's use of science provides the framework for the novel. Filled with encounters between the submarine *Nautilus* and various "monsters" of the deep sea, including creatures that appeared to be dugongs and manatees, the book captured both the fear and excitement of deep-sea exploration. Victorians' fascination and terror of monsters was vividly portrayed in the description of the attack on the *Nautilus* by a giant squid. It is difficult to know where the science ends and the fiction begins. The narrator Aronnax, a professor of Natural History at the Museum of Paris was asked to join the expedition of the *Abraham Lincoln* in pursuit of the monster. As a man of science, he assured the reader his account was based on logic and cool hard analysis, not wild exaggeration. He pointed out that leading naturalists such as Cuvier and Lacepede would never had admitted that such a creature existed if they had not actually seen it with their own eyes. Numerous sightings at a variety of times and places confirmed that the monster did indeed exist. "As to classifying it in the realm of the false, the idea was out of the question." Adopting the style of a ship's log, giving dates, longitudes and latitudes of the *Abraham Lincoln*'s position in its hunt for the monster furthered the nonfictional aspect of the novel. Conseil, Aronnax's servant mirrored his master's rational disposition with his incredible memory and ability to classify knowledge. Ned Land was a harpooner, afraid of neither death nor the unknown. He also was loath to accept a supernatural explanation when a natural one could be found.

The monster was neither natural nor supernatural, but a submarine powered by electricity. The *Nautilus* provided an additional forum for Verne to expound on the developments in science. His detailed description of the power of locomotion, underwater pressure, and renewal of air was designed to convince even the most sophisticated reader that such a ship might really exist. But beneath the surface story of the submarine is a subtext that provides additional insight into the Victorian psyche and their fascination with the sea serpent.

Exquisitely outfitted, with every imaginable item of luxury, the *Nautilus* was a man-made monster, a work of genius, capable of amazing and also deadly feats. Its creator, Captain Nemo was "a perfect archangel of hatred" who made the narrator and his two shipmates eternal prisoners as they explored the sea and discovered real monsters of the deep. Thus, sea monsters were portrayed as both fictional and real, but as embodied in the *Nautilus* itself, the sea monster represented both the utopian and sinister possibilities that advances in science and technology bring to society. The novel tapped into the Victorian anxiety over scientific findings that suggested we live in a materialistic world, devoid of spiritual meaning. Sea serpents as a genuine, natural relic from the past represented to Victorians a sense of continuity in a world that for many was changing far too rapidly. At the same time the very elusiveness of the serpent left room for the imagination and gave hope to the idea that there would always be mysteries that would resist scientific explanation. Yet more and more mysteries were becoming unraveled as science extended its reach into topics that were previously under the purview of philosophers and theologians.

One such mystery was the nature of mind. Phrenology offered the possibility of a scientific understanding of mind and behavior and attracted an enormous following, spawning a social movement that tapped into both the fears and hopes of Victorian society. However, just as with the sea serpent, not only would the quality of evidence be mired in controversy, but phrenology also became enmeshed in boundary disputes as various factions within the emerging scientific community scrambled to gain a monopoly on who determined what was considered legitimate science.

3

Franz Gall, Johann Spurzheim, George Combe, and Phrenology

A Science for Everyone

We are reading Gall's "Anatomie et Physiologie du Cerveau" and Carpenter's "Comparative Physiology" aloud in the evenings.

—George Eliot, *George Eliot's Life as Related in Her Letters and Journals*

PHRENOLOGY IS OFTEN cited as *the* classic example of pseudoscience with its practitioners dismissed as quacks. Lumped together with physiognomy, graphology, and palmistry, it has been called "psychology's great *faux pas.*" One reads about it in books such as *Fads and Fallacies in the Name of Science.*[1] However, George Eliot's letter belies such a simplistic dismissal and illustrates how closely phrenology was associated with conventional physiology. In actuality, phrenology played a crucial role in furthering naturalistic approaches to the study of mind and behavior.[2] It attracted followers from all strata of society—from Prince Albert who consulted George Combe[3] about his eldest son's cranial bumps, to George Holyoake, an Owenite, secularist, and leader of the cooperative movement in Britain. Evolutionists Alfred Wallace, Herbert Spencer, and Robert Chambers were convinced of phrenology's worth along with other prominent Victorians including John Stuart Mill, George

Lewes, Harriet Martineau, Samuel Coleridge, and Auguste Comte. Phrenological ideas also found their way into the writings of Charlotte Bronte, Edgar Allan Poe, Walt Whitman, Nathaniel Hawthorne, Herman Melville, and Sir Arthur Conan Doyle. Issues of professionalization, class differences, religion, and politics as well as philosophy were inextricably intertwined in the phrenological debates. Phrenology in its historical context, in fact, exemplifies how difficult it is to draw boundaries between science, pseudoscience and nonscience.

Further complicating the evaluation of phrenology is that the term phrenology encompasses several different programs. The supporters of the cerebral physiology of Franz Gall were often attacked for espousing ideas that find virtually unanimous support among modern-day scientists. Johann Spurzheim promulgated the "science" of phrenology that George Combe used to build a social reform movement. Phrenology took on a life of its own far removed from the original ideas of Gall as people flocked to fairs and carnivals to have the bumps on their head "read." At the same time, highly technical discussions were occurring about the merits and drawbacks of phrenology in prestigious medical journals such as *Lancet* and the *London Medical and Physical Journal*. In a classic understatement one article pointed out, "understanding the brain was a very important question" and that phrenology had the "novelty of being both ingenious and amusing."[4] Although many of the ideas of phrenology seem patently ridiculous to us today, the basic underpinnings as developed by Gall remain the basis for our understanding of mental phenomena. These included that the brain is the organ of the mind and that the brain is localized for specific functions. In Gall's attempt to form a consistent *biological* theory of brain function, he faced fundamental problems in not just physiology, but ontogeny, epistemology, and psychology as well. Gall never used the term phrenology; rather, he referred to his own work as cerebral physiology. But since phrenology is most often associated with the ideas of Gall, his theories will be the starting point for an exploration of a doctrine that initially involved primarily physicians and scientists, but quickly spread to the larger public, engendering a movement addressing questions in philosophy, religion, education, and social policy.

The Organology of Franz Gall

Franz Joseph Gall was born on March 9, 1758, in Tiefenbrum, Baden, Germany, the son of a merchant. Both of his parents were devout Roman Catholics and they hoped their son would make his life in the church. However, this was not to be. His early passion for science was followed by a strong taste for women and money and his marriage did not keep him from having mistresses or fathering an illegitimate son. His disregard of accepted standards of moral behavior combined with his unorthodox scientific ideas insured that a cloud of controversy surrounded him wherever he went.

Gall began his education under the tutelage of his uncle, a priest and in the Baden schools, but then went on to study medicine in Strasbourg. He moved to Vienna and received his MD degree in 1785 and set up a very successful practice. While practicing medicine, he decided to formally investigate ideas that had interested him for some time. Gall was influenced by physiognomy: "the act of judging character and disposition from the features of the face or form and lineaments of the body generally" (*Oxford Dictionary*). Such ideas were in great vogue, mainly due to the writings and lectures of the Swiss priest Johann Caspar Lavater (1741–1801). Lavater was also a student of medicine and had maintained, as Gall was later to independently claim, that the external form of the brain imprinted itself on the internal surface of the skull as well as modeling the contours of the external surface.

As a boy Gall had noticed that his siblings and friends all exhibited distinct talents and dispositions. One might excel in penmanship, another in arithmetic while others had a great ability to learn languages or natural history. Students who had good memories often had bulging eyes. In college, he observed that there was often a correlation between the shape of the head and particular abilities such as in painting or music. Since each physiological function had its own organ as did each of the external senses, was it not reasonable that the various talents, instincts, and propensities would have their own organ? Thus, he started to look to the head for signs of the moral sentiments. However, Gall was very clear in pointing out that the skull was not the cause of these differences between people, but rather the brain.

By 1791, Gall had formulated his basic ideas of brain function. These can be summarized as follows:

1. The brain is the organ of the mind.
2. The brain is not a homogenous unity, but an aggregate of mental organs with specific functions.
3. The cerebral organs are topographically localized.
4. Other things being equal, the relative size of any particular mental organ is indicative of the power or strength of that organ.
5. Since the skull ossifies over the brain during infant development, external craniological means could be used to diagnose the internal states of the mental characters.

Throughout the 1790s, Gall made systematic observations, mainly of inmates in prisons and mental asylums, trying to find correlations between the external structures of the head and general powers such as memory, judgment, and imagination. However, the correlation between these characteristics and the shape of the skull was very poor. He turned to more specific traits, for example a talent in music, painting, or mathematics and claimed to be more successful. He eventually came up with twenty-seven organs or faculties of the brain (see figure 3.1). Gall divided the faculties into two major groups. The first were feelings that had two subcategories: propensities and sentiments. Propensities also existed in the lower animals and were located at the base of the brain. They included such faculties as adhesiveness, which was the tendency to live in communities and amativeness or physical love. Sentiments, however, were found only in humans and they included the faculties of imagination, conscientiousness, and firmness. These nobler sentiments were located near the crown of the head. The second major group consisted of the intellectual faculties that were subdivided into the "knowing" faculties and the "reflecting" faculties. Knowing faculties such as form, size, color, and number were how individuals used sense perceptions to learn about the external world. All organisms had such faculties to some degree. It was the higher order reflecting faculties that set human beings off from their animal cousins. These faculties included comparison, wit, and language and were located in the front of the skull. Since the size of a particular region of the brain was a measure of the development of the faculty that was located there, a person with a large skull with a high full brow had highly developed faculties of reflection, or at least the potential of such development. However, this was not a simple craniology that Gall espoused. Turning away from physiognomy, he investigated what he called cerebral physiology and the distinction is important. According to the principles of physiognomy, a person's features *express* his character. Gall disagreed. The body is not the expression of the character. Rather "the object of my researches is the brain. The cranium is only a faithful cast of the external surface of the brain and is consequently, but a minor part of the principle object."[5] Furthermore, physiognomy never endeavored "to tell what were the original propensities of man, much less to indicate the simple fundamental faculties of our nature."[6] Gall was claiming that human character and intellect were the result of the combined functions of many organs localized on the surface of the brain. Mental processes had a purely materialistic basis that could be identified. The "organs" could increase or decrease in size according to the amount of specific faculties the individual possessed. Thus, people who had good memories had bulging eyes because the "organ" of language that included memory was located on the orbital surface of the frontal lobe. An aggressive person had cranial protuberances behind the ears because, according to Gall, this was the location of the organs of combativeness. In a similar vein, the bump that correlated to the faculty of acquisitiveness tended to be quite prominent in pickpockets.

Gall continued to develop his ideas and presented them in public lectures and demonstrations in Vienna. Attending one such lecture was Johann Spurzheim, who was studying medicine in Vienna and became his assistant and collaborator. Tracing the various nerve fibers, Gall saw that they did not converge at some single point, which suggested that no single point existed that could be identified as the seat of the soul. Heresy! Gall was accused of corrupting morals and endangering religion. The emperor of Austria in a personal letter to Gall banned his lectures, claiming that they promoted materialism, immorality, and atheism. Repeated appeals and petitions were to

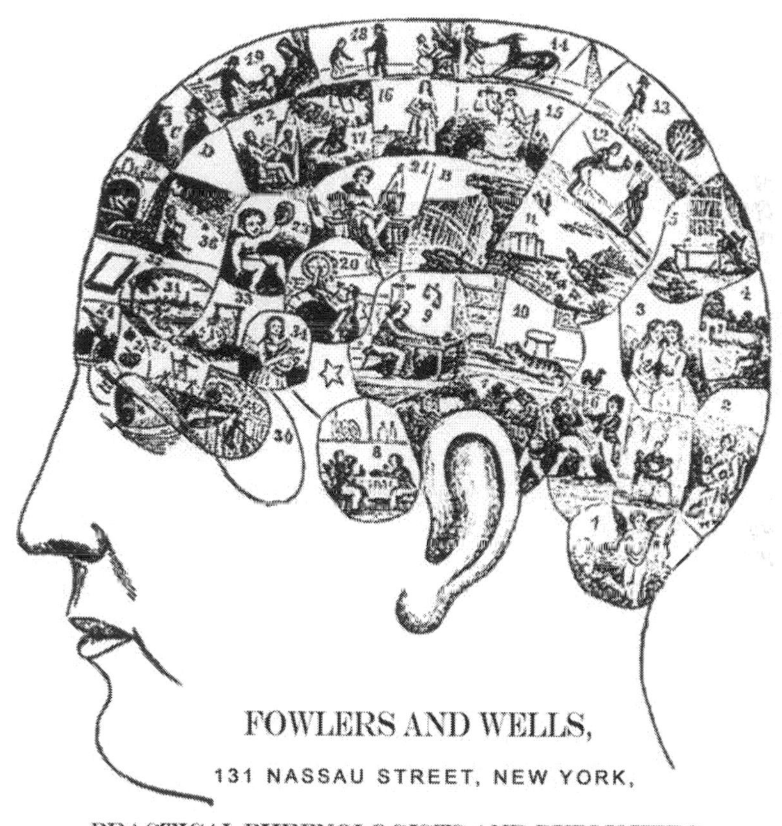

FIGURE 3.1. The Organs of Phrenology. From George Combe, *The Constitution of Man Considered in Relation to External Objects* (New York: Fowlers and Wells, 1848).

no avail and in 1805 Gall left Vienna with Spurzheim. They traveled through northern Europe visiting hospitals, prisons, and insane asylums to gather evidence and demonstrate their doctrines. Their travels were highly successful, if also enmeshed in controversy. Gall's doctrines as well as his character were often attacked. Nevertheless, by the time they settled in Paris in 1807 Gall was quite famous. Both men continued to write, lecture, and develop their theories.

In 1813 Spurzheim left for England to spread the doctrine, but it was also the beginning of a schism between the two men. Publishing under his own name and deviating from many of Gall's concepts, he gave a new name to the whole system referring to it as "phrenology," which was derived from the Greek work "phren" for mind.[7] The term phrenology became associated with both men's work in spite of Gall's disagreement with many of Spurzheim's later claims. Gall did not use the term himself, referring to his ideas as cerebral physiology.

Although Gall had little direct experimental evidence for his claims, as was previously pointed out, this was not why people initially objected to his doctrine. Rather, phrenology had decidedly materialistic implications. For Gall, cerebral physiology was simply the science of the mind. However, in it are the underpinnings of his religious views. Gall has been rightly called a deist as he specifically disavowed atheism, claiming the idea of atheism was absurd. But he was indifferent to particular religious dogmas or creeds. He considered the soul of humans as one and the same as the brain, but he did not want to become involved with discussions about the immortality of the soul because "this does not concern the physiologist directly." Unlike many people within the tradition of natural theology who wanted to use the findings of science to prove design, he took design for granted. The makeup of a particular human's character was God given; the human race was neither good nor bad, but both. Humans were endowed with such positive qualities as benevolence and religious feelings, but man will also steal and murder. He thought there would always be wars since these acts were founded in humankind's natural desires. For Gall a strict harmony between moral and physical laws existed. Both had their source in God and there cannot be any discrepancy between them. All man does, all he knows is due to God, and the cerebral organs are the intermediate instruments that he uses. "God and the brain, nothing but God and the brain!"[8]

Gall included religion as one of his twenty-seven organs and claimed that it was additional proof for the existence of God. Nevertheless, Gall's doctrines were denounced as atheistic. His books were placed on the Index and he was denied a religious burial. Physiognomy could accommodate a soul by claiming that the shape of the body reflects the soul just as man was created in God's own image. Gall repudiated such analogies, maintaining instead that "the convolutions of the brain must be recognized as the parts where instincts, sentiments, penchants, talents, and in general the moral and intellectual forces are exercised."[9] However, claiming that the brain is the means by which thought is expressed is not the same as claiming that the brain manufactures thought or that thought consists of the material substance of the brain. Gall believed that it was only possible to investigate what can be experienced in the physical realm. "We carry out our researches neither on the inanimate body nor on the soul alone, but on the living man, the result of the union of soul and body."[10]

Gall hoped to understand the function of the brain by relying on the methods of anatomical research, but he denied being a materialist. Humans were intellectual and moral beings, but he did not claim that the brain or the divisions of the brain produced these qualities. Rather, the cerebral organs were the instruments through which these basic qualities expressed themselves. Nevertheless, his methodology certainly had materialist underpinnings. It was only possible to study the moral, intellectual, and spiritual nature of humankind because those psychic qualities had a direct relationship to the organs of the brain. The qualities of the mind were not just an accidental product of the sensory perception of external entities.

Gall was not the first person to argue that the brain was the organ of the mind or that mental phenomena should be investigated in terms of physiological function. Nevertheless, anatomists and physiologists had studied neither the functions of the brain nor how they might be localized in detail. The philosopher-psychologists considered other problems of mind far more important such as trying to locate the seat of the soul. In the 1600s, Descartes situated the soul in the pineal gland of the brain, but he did not identify the brain with the mind. He instead argued for a mind/body dualism with the pineal gland merely being the point at which the mind affects the flow of animal spirits and the immaterial soul made contact with the physical brain.

Understanding the nature of mind had been considered fundamentally a philosophical problem. Phrenology played a prominent role in psychology eventually breaking from philosophy and in its struggle to establish itself as a true experimental science—something that still plagues it today. At the heart of Gall's theory lay questions regarding the relationship of the body to the mind and the problem of primary and secondary qualities of the mind. Gall had hoped that by taking a biological approach he could resolve these problems of a dualistic ontology that was the legacy from Descartes.[11]

Gall distanced himself from the prevailing sensationalist theories of Étienne Condillac and its variants such as "idealogy," in spite of them having many ideas that resonated with his own. Condillac had been primarily concerned with the sensations of the external senses. Others argued that all intellectual phenomena were modifications of feeling that were merely awareness of four basic kinds of phenomena: sensibility, memory, judgment, and will or desire. They went beyond Condillac in recognizing that internal sensations were responsible for instincts and passions. However, they believed that the emotions have their seat in the internal organs. Gall made explicit the relationship of all psychic phenomena including instincts and passions to the central nervous system. Nevertheless, the ideologues were treating mental phenomena in physiological terms and provided a climate that was congenial to Gall's ideas.

Gall's work along with that of the ideologues was subject to various attacks on both religious and philosophical grounds. Claiming that such concepts as freewill were the result of other more primary faculties of the brain led to charges of materialism and atheism. His idea of dividing the brain into distinct organs was philosophically abhorrent to followers of Descartes who believed in the duality of the mind and the body, but also that the mind was indivisible. Moral philosophers in the Scottish tradition also rejected the phrenological approach to understanding the nature of mind. Instead they advocated "reflection," basically a method of introspection. Only by systematically examining one's own mind at work, by looking inward, would it be possible to unlock the secrets of the mind.

Gall had many ideas in common with the sensualists. He regarded sensibility as the most general property of the nervous system. However, in the higher animals and in humans, the whole nervous system was under control of the brain and sensibility cannot be distinguished from sensation, that is, the perception of a stimulus. For Gall, sensibility without consciousness was a contradiction in terms. Therefore, the many reactions that go on in our body without us being aware of them had to be explained by irritability. Gall, by claiming that the brain was the primary organ responsible for sensibility makes it the unique organ of all higher functions, which included instinctive urges and passions. This idea had important ramifications, particularly in regard to psychiatry, and for policy recommendations in a wide range of social issues from education to the treatment of criminals.

In spite of the similarity of Gall's ideas to the sensualists, he was critical of their theories in significant ways. First of all, he denied Condillac's assertion that we receive all our ideas and knowledge through the external senses alone. Such a supposition could not explain either the observed differences between the talents and propensities of individuals or differences between species. Character and personality could not be explained by experience alone. Rather, what we think and desire is a combination of both the internal organization of the brain and the impressions from the outside world. Traits such as memory, judgment, and desire were common to all primary qualities and functions of the brain. However, what were these primary qualities and their functions? This was the question that Gall regarded as most fundamental and attempted to answer when he described his twenty-seven faculties and their location. Gall observed his fellow humans as well as animals and compared their various instincts, talents, and deficiencies, trying to locate them in the brain. He shifted the investigation of the mind from an introspective and psychological analysis of sensation to a descriptive and physiological investigation. In doing so he came much closer to carrying out the ideologue's program than they themselves had done, and a naturalistic, materialistic approach to the study of mind and brain could no longer be ignored. Furthermore, both he and Spurzheim insisted that the anatomy of the brain had to be studied from a physiological point of view, whereas traditional anatomists advocated a purely structural approach. Cerebral physiology promised to explain the mysteries of personality and character in a way that went against the prevailing views of human nature espoused from both the pulpit and the university lectern.

Gall pioneered an approach in studying the brain that has been invaluable. Unlike most of his contemporaries, Gall did not dissect the brain by slicing it from top to bottom that would destroy the connections of its various parts. Instead, he followed the brain's own structural organization and attempted to trace the course of the nerve fibers and their relationship to the various gray masses from which they were derived and in which he believed that they ended. Pierre Flourens, Gall's most persistent and successful critic, admitted in 1863, by which time most of Gall's most popular theories had been essentially totally discredited, that when he first saw Gall dissect a brain he felt as if he had never seen it before and called Gall "the author of the true anatomy of the brain." Gall made important contributions to understanding the organization of the cerebral cortex that helped discredit humoral and glandular conceptions of brain function. Both his and Spurzheim's work provided an important stimulus to neuroanatomical research, which ironically contributed to the downfall of many of his detailed claims.

In spite of Gall's extensive and detailed dissection of brains, his most important ideas were *not* an outgrowth of this research. His belief in the innateness of the moral and intellectual faculties and that the brain consisted of a plurality of the cerebral organs, which were essentially independent of one another, came prior to his undertaking of detailed dissections. Furthermore, his findings did not lead to significant modifications of his theory. Although Gall argued that any theory of brain function had to be compatible with its structure, he also maintained that knowledge of function had always preceded that of structure. All his physiological studies had been made without knowledge of structure, and he brazenly claimed that they could have existed for ages in the absence of any understanding about the anatomy of the brain. There is an irony here, since he was making an explicit claim about the relationship of structure to function; and it is these claims that brought him into conflict with the prevailing philosophical and religious views. It was precisely because he maintained that feelings, instincts, sentiments, and intellects were localized in a specific part of the brain that he was accused of promoting materialistic and atheistic doctrines. Yet neither his nor Spurzheim's anatomic researches provided compelling evidence for these "organs." With one exception, none of his localizations of cortical function has been held up by later research. Nevertheless, the fundamental idea that different parts of the brain had different physiological and, therefore, psychophysiological functions turned out to be correct and has provided an enormously fruitful research strategy. His innovative approach to brain dissection laid the foundation for the detailed kinds of exploration of the brain that characterized the research of the next century.

Gall was a profound thinker, but several serious methodological problems existed in his work. First, his categorization of the faculties was prior to and independent of the detailed dissections that he did with the help of Spurzheim. While Gall reiterated again and again that the brain was the organ of the mind, and that function and structure were intimately related, in practice he did not carry out this program. Instead, he insisted that function preceded structure and his anatomical work only played a confirmatory role in the elucidation of brain functions. He argued that simple dissection, neuropathological studies, ablation experiments, and comparative anatomy had not led to and never could lead to knowledge of the function of the brain. How did he come to what seems to a modern reader such a bizarre conclusion?

According to Gall, morphology of the brain by itself was meaningless. What was the point of dissecting the brain unless one knew what she was looking for? Previous attempts at cerebral localization had failed because no attempt had been made to find the fundamental, primitive faculties. Only observing animal behavior and the moral and intellectual characteristics of individuals in nature and society could discover these faculties. Once they had been discovered, then anatomical, pathological, and clinical studies along with comparative anatomy were useful to confirm the psychological findings. When Gall did attempt to discover where particular functions were located in the brain, he confined his investigations to the cortex. Although he maintained that it was the brain he was writing about, he relied primarily on observations of the cranium because he assumed that the skull really was a faithful cast of the underlying brain. In most of his writings it is not clear whether he actually was examining the brain or whether he had inferred the existence of a particular organ from the protuberances of the overlying cranium. His claim that pride in humans and the ability of certain birds to fly very high were due to similar cranial protuberances suggests that simple craniology was the basis of much of his evidence. He continually asserted that when he compared the cranial findings against actual direct observation of the brain that the findings based on the cranium perfectly correlated. However, it is unlikely that he had made direct observations because it would have been impossible to do so in most cases. He admitted that his organs did not divide up according to the convolutions of the brain and that he could not localize precisely the extent of each organ. In actuality Gall was far more interested in describing the nature of the functions of his organs than in their localization. He also was not particularly interested in developing the delicate neuroanatomical techniques that would be necessary to accurately map the localizations of brain functions. Unfortunately, most of phrenology's advocates had even less interest than he did, as phrenology degenerated into "bumpology." Gall's interest in function was also quite narrow. He described his work as cerebral physiology, but his notion of physiology was quite different from how it is described today. His curiosity only went as far as determining what the brain's functions are, but not how it actually carried out those functions. To be fair, the state of physiology in Gall's time was rather primitive. As he rightly pointed out, even in the cases of most other organs, not much was known on how they actually worked. Knowledge of the organs for intellect and moral qualities was essentially in the same position as physiologists' understanding of kidney function. As our knowledge of the brain has increased, trying to relate the fine structure and physiology of the brain to actual behavior has become harder, not easier. Gall's goal was much more limited—to identify a particular quality or faculty and then use that knowledge to discover the location of the organs that were responsible. Even that task has been enormously difficult and still has not been accomplished.

Understanding brain function would lead to the discovery of "the organs of the moral and intellectual faculties of man," which Gall regarded as the most important aspect of brain physiology.[12] In spite of the problems with Gall's research, his ideas laid the groundwork for examining brain development from an evolutionary perspective. Indeed, comparing Gall's methods and evidence with that of Darwin's illustrates the blurred edges between science, marginal science, and pseudoscience.

The Phrenological Connection to Evolutionary Theory

Today, Darwin is still read and cited by every serious student of biology and psychology. Gall is relegated to a passing acknowledgment in histories of neurology and psychology. Yet Darwin's debt to phrenology was enormous. Victorians had already been exposed to the idea that animals shared many of the same mental qualities as humans and that their brains were organized in a similar manner in 1844 with the publication of the

Vestiges of the Natural History of Creation. The author explicitly acknowledged the importance of Gall in the development of his ideas.[13] Darwin later used many of the same kinds of evidence as Gall to argue that humans were not an exception to his theory, and as we will see, striking similarities exist between passages in *Vestiges* and Darwin's *The Descent of Man and the Selection of Sex in Relation to Man* (1871). But before explicitly comparing *Vestiges* to *The Descent of Man*, it is worth noting the similarities between Gall and Darwin not only in terms of the types of evidence they used, but also in the logical structure of their arguments in support of their respective ideas.

Both Darwin and Gall collected enormous amounts of data in favor of their theories. Both men assumed that the more observations they collected, the greater the likelihood that their theories were correct. A significant difference (and one that ultimately contributes to phrenology being labeled pseudoscience) is that generally the quality of the evidence was higher in the evolutionary studies. Nevertheless, Darwin's evidence for the theory of natural selection consisted of naturalistic observations combined with a great deal of anecdotal evidence systematically collected. In *The Descent of Man* he used comparative embryology and anatomy to illustrate that human brains were organized similarly to animal brains, particularly the higher apes, but the anatomical evidence was just the groundwork for Darwin to lay out his theory of the origin of a moral sense. What is worth emphasizing is that anecdotal evidence of animal behavior was *the* primary kind of evidence that Darwin used to support his idea that the moral sense evolved by natural selection. The beginnings of a moral sense could be seen in the behavior of the lower animals. This is not meant to be a criticism of Darwin. Rather, it was the standard of evidence at the time. The kind of statistical methods for analysis that are used in present-day studies were not available. Logically, Darwin's theory was in the same position as phrenology for much of the nineteenth century.[14]

Darwin's theory provided a causal explanation for an enormous variety of observations. He did not think that the correlations he observed to provide evidence for his theory were merely chance events. Gall also did not think that his earliest observations of the correlation between bulging eyes and good memories were merely a coincidence. Gall and the followers of phrenology maintained that the correlation between cranial prominences and particular behaviors represented a causal relationship. Correlation does not necessarily mean causation, but Gall had collected a variety of evidence to support his claim.

First, when Gall noted that a person or animal had an unusual talent or propensity, he tried to find out if this was due to nature, that is, what he considered a fundamental faculty. Generally a propensity was considered fundamental if it expressed itself independently of the other characteristics of the individual or species. Gall searched for a correlation between the talent or propensity and a cranial protuberance, collecting and comparing as many correlations as he could in both animals and humans. He then examined individuals who either had a moderate deficiency or were lacking entirely in this propensity to see if the cranial protuberance was also lacking. By 1802 Gall had collected over 300 skulls and had 120 head casts of people with special talents and propensities. Paintings of famous people, quotations of famous authors, and anecdotes from any source all contributed to his evidence in favor of correlations between particular propensities and skull protuberances. He claimed the correlations were excellent in healthy and middle-aged brains.

Gall was actively seeking confirmations, however, and when he found them, he had no real standards for comparison. Everything was used, from reports in the press to literary quotations, and there was no attempt to discriminate between them. Furthermore, he had numerous ways of explaining away apparent contradictions to his theory. If a propensity was expressed when the protuberance was small or absent when it was large, this could be due to the interactions of the other organs, due to disease, or due to habit or education. For example, Gall suggested that the bulging eyes in a boy, but whose memory was only average was due to rickets or hydrocephalus. Spurzheim also was undaunted by contradictory evidence. When it was pointed out that in Descartes' skull the superior and anterior regions of the forehead where the organs of rationality were localized were quite small, Spurzheim retorted that Descartes was not all that great a thinker![15] However, reinforcing observations and evidence that confirmed their views while explaining away cases that did not, was not a practice unique to phrenologists. It is a major source of cognitive bias and remains one of the chief sources of error in science that still has not been successfully rooted out by the development of controlled and double-blind experiments.

Nevertheless, much of what was cited as evidence in favor of phrenological doctrine were descriptions of people whose character was well known and then the appropriate "organs" were stated to be correspondingly large or small. The description of Charles Darwin in the *American Phrenological Journal* is typical of the types of articles supporting a phrenological analysis of character. According to phrenology, Darwin's character was so clearly evident in the organization of his brain that:

> It scarcely needs an experienced physiognomist to read it. The towering crown indicates positiveness, self-reliance, decision, independence. Intellectually, we would regard him as the ready observer, the facile inquirer, the keen investigator. His well-marked reflective organs evince the close and profound analyst rather than the merely speculative thinker; the weigher and adapter of

PORTRAIT OF CHARLES DARWIN, THE NATURALIST.

FIGURE 3.2. Portrait of Charles Darwin. From *American Phrenological Journal and Life Illustrated* (1868).

> facts rather than the theorist. . . . His appreciation of mere probability is very slight; his organ of mere belief is very weak. We cannot give him credit for much Veneration, Hope, or Spirituality, and his lack of these organs tends to the sharpening of his practical and utilitarian views of things.[16]

Darwin demanded "facts before hypotheses, with substantial premises before ratiocination." This was an accurate portrayal of what was known about Darwin, but where were these organs of Veneration, Hope or Belief even in a religious person? The reader must take it on faith that Darwin's skull does exhibit the appropriate protuberances reflecting the underlying organs (see figure 3.2). Darwin might have found this phrenological analysis of his character foolish, but he would also find that Gall's doctrine was far more than a system of lumps and bumps.

In spite of what seems to be quite poor evidence by modern-day standards, an important aspect of Gall's evidence that sets it off from the practice of physiognomy and that of later phrenologists was that Gall did not confine his observations to humans alone. Like Darwin and pre-Darwin, Gall argued that a very close relationship between humans and animals existed. Many of the lower animals shared similar psychic qualities that reflected the similar organization of the cerebral organs. He claimed that humans and animals shared nineteen of the twenty-seven fundamental faculties. These were (1) the sexual instinct, (2) love of progeny, (3) attachments, instincts of (4) self-defense, (5) killing, and (6) cunning, (7) the desire to possess things, (8) pride, (9) vanity, (10) circumspection, (11) memory for facts, (12) sense of locality, (13) memory for persons, (14) memory for words, (15) a sense of language, (16) sense of relations of color, (17) of musical tones, (18) of numbers, and even (19) a mechanical sense existed in animals to some degree. The difference between humans and animals was only of degree, not of essence. This was precisely the same argument that Darwin used in *The Descent of Man* in 1871.

Gall disagreed with the idea than animals were just machines, devoid of feelings or of moral and intellectual principles. He did not accept the prevailing view that humans were unique because they alone possessed an immaterial substance and, therefore, only humans had freewill and reason. He had no use for the idea that humans alone had a soul that acted independently of the body. He still thought humans were superior, but the question remained what made them different from animals? For Gall, the answer would be found in the comparative anatomy and physiology of the brain. The functions of the lower anterior part of the brain were much more developed in humans than in animals. Furthermore, the upper anterior part of the human brain was virtually nonexistent in animals. Thus, it was in this area of the brain that one would find the organs responsible for specifically human qualities. These included organs of comparative sagacity, spirit of metaphysics, wit, poetic talent, goodness, initiative, religious sentiment, and firmness. However, he claimed that some of these faculties such as goodness could be observed in the lower animals, and thus, he was an extremely important thinker in breaking down the idea that some unbridgeable chasm separated humans from animals. Just like Darwin, he maintained that the difference in the moral and intellectual qualities between animals and humans was one of degree only. Although humans possessed certain qualities that animals lack, they were not so different from those faculties shared by both. In Gall's scheme no special organ of consciousness, intellect, reason or will existed. There were many faculties that made up "intellect." Gall did not make a clear separation between instinct and intellect. Rather, he claimed that there existed as many "intellects" as there were qualities and these qualities ranged from mere instinct to highly conscious and intelligent faculties. Interaction and cooperation between the various faculties resulted in very complicated mental acts that explained the phenomena of reason and freewill. Gall's work was especially important as it paved the way for an evolutionary account of the nervous system. Reading Gall it is hard not to think that Darwin must have read and been influenced by him, but clear documentation of this has not been found. Instead, Darwin appears to have gotten many of his ideas about animal minds from the natural theology literature.[17]

Darwin did not immediately apply his principle of natural selection to behavior. Although Paley's *Natural Theology* had been an important inspiration to his speculations, it was also a formidable obstacle—at least initially. Animals performed a variety of complex behaviors such as nest building, food gathering, and migration without any prior experience or learning. How could a newly mated bird know that the purpose of nest building was for incubating eggs it hadn't seen? Paley argued that all such complex behaviors were instinctive, part of the Deity's grand design. God had bestowed reason on only one chosen species. However, not all natural theologians agreed with Paley. For some, a cat learning to open a door from watching people provided sufficient proof of rationality in animals. Animals were capable of modifying their instinctive acts to adapt to changing circumstances, demonstrating that there was a continuum between the instinctive behavior of animals and human rationality. Although behavior in insects was primarily instinctive, they were also capable of reason. They could learn from experience, employ memory, and had complex means of communication that enabled them to change their behavior according to circumstances. Nevertheless, the natural theologians concluded that such behavior as well as the order and complexity of the world, especially the adaptation of each individual to its environment could only be the result of an intelligent designer.

Darwin eventually came to a very different conclusion. He not only came to believe that his principle of natural selection could explain how species change through time, but that this principle could also be used to explain the origin of complex behaviors, including the origin of a moral sense. No divine intervention was necessary. Darwin would have an uphill battle to convince people of the truth of such radical ideas. However, Victorians had already been debating some of the most controversial aspects of Darwin's ideas many years before the publication of *The Origin* (even if couched in the language of natural theology), having been exposed to them in *Vestiges of the Natural History of Creation.*

According to *Vestiges,* from the simplest of forms life progressively changed, eventually giving rise to organisms such as human beings with a rich and complex mental life. Darwin read *Vestiges* carefully and was paying close attention to the reaction it engendered. He thought that the author's "geology strikes me as bad, and his zoology far worse."[18] But to himself, he had to acknowledge that the anonymous author had used many of the same kinds of evidence—embryology, the fossil record, classification and comparative anatomy—that he had gathered for his own theory of species change through time. Darwin had much more in common with Mr. Vestiges than he was willing to admit in public.

The anonymous author was Robert Chambers, a Scottish journalist, publisher and a thoroughgoing phrenologist. *Chamber's Journal* was filled with articles promulgating phrenological doctrine, and people speculated that he might be the anonymous author. Chambers had also drawn heavily on the natural theology literature in developing his ideas, and references to them were made throughout the first part of the book. But rather than the theologians, it was Gall whom he explicitly acknowledged as being the crucial influence in his thinking about the mental life of animals. Darwin, already primed by his reading of the natural theology literature would find further reinforcement for his ideas in *Vestiges.*

The similarity between particular passages in *The Descent of Man* and *Vestiges* as well as Darwin's whole line of argument is quite striking. Compare the following:

> CHAMBERS: The difference between mind in the lower animals and in man is a difference in degree only; it is not a specific difference.[19]
>
> DARWIN: The difference in mind between man and the higher animals, great as it is, is certainly one of degree and not of kind.[20]

and concerning the higher animals, especially the primates

> CHAMBERS: We see animals capable of affection, jealousy, and envy. We see them quarrel. . . . We see them liable to flattery, inflated with pride, and dejected by shame.[21]
>
> DARWIN: A dog carrying a basket for his master exhibits in a high degree self-complacency or pride. There can be no doubt that a dog feels shame as distinct from fear. . . . Several observers have stated that monkeys certainly dislike being laughed at.[22]

Who could deny the similarities between themselves and the lower animals as they read in *Vestiges* about horses who wanting a shoe on their own accord returned to the shop where they had been shod before, and cats who figured out how to be let out of a room by pulling on a latch or ringing a bell.[23] Darwin used a similar strategy in *The Descent of Man* to illustrate the continuum between man and beast. "Sympathy leads a courageous dog to fly at any one who strikes his master. . . . All have the same senses, intuition and sensations—similar passions, affections, and emotions: they feel wonder and curiosity; they possess the same faculties of imitation, attention, memory, imagination, and reason, though in very different degrees"(see figure 3.3).[24]

Chambers drew a parallel between the education of humans and the domestication of animals. "Between a young unbroken horse, and a trained one there is all the difference which exists between a wild youth reared at his own discretion in the country, and the same person when he has been toned down by long exposure to the influences of refined society."[25] Darwin also discussed the influence of domestication on animal behavior. Dogs were descended from wolves and as a result they have "progressed in certain moral qualities such as in affection, trust-worthiness, temper, and probably in general intelligence."[26] Like Chambers, Darwin used animal stories to provide evidence in favor of his theory, and both men undoubtedly culled many of the specific examples from the writings of the natural theologians.

In building his case for the similarity of animals to humans, Chambers credited Gall as being crucial to his thinking on the matter. Gall had been the only philosopher who had developed a system of mind that was founded on nature. He had identified particular parts of the brain referred to as faculties that were associated with various capabilities such as imitation and wonder. There were parts of the brain for the moral feelings of benevolence, conscientiousness, and veneration. Chambers, like Gall, believed that these capabilities were far more developed in humans. Nevertheless, he argued that they existed to some degree in the lower animals. Many of the human faculties that Gall had identified existed as instincts in animals, but represented an early stage of development. The cell formation of the bee, the house building of ants and beavers, and the web spinning of spiders were for Chambers just primitive exercises of the faculty of constructiveness. In humans these instincts led to the arts of the weaver, upholsterer, architect, and mechanist. The storing of provisions by ants was indicative of the faculty of acquisitiveness that in humans created rich men and misers. Many insects provided for protection and sustenance of young they would never see. According to phrenological doctrine, such behavior was just a less developed form of the faculty of philoprogentiveness (love of progeny).[27] Mr. Vestiges was not the only person making such suggestions. Darwin described a monkey who after breaking open nuts with a stone, hid the stone in the straw and wouldn't let any other monkey use it. "Here, then we have the idea of property; but this idea is common to every dog with a bone, and to most or all birds with their nests."[28]

Darwin also described various characteristics and abilities in animals that were commonly thought to be unique to humans and he referred to these traits as faculties. They included self-consciousness, individuality, and having a sense of beauty. He had numerous examples of animals that displayed imagination, wonder, and curiosity as well as sympathy and affection. Gall claimed there was an organ of religiosity. Darwin recognized that the feeling of religious devotion was highly complex, but maintained that the beginnings of this state of mind could be seen in "the deep love of a dog for his master, associated with complete submission, some fear, and perhaps some other feelings."[29]

FIGURE 3.3. Expression of Emotion in a Dog. *Top:* Dog showing hostile intent to another dog. *Bottom:* Dog in a humble and affectionate state of mind. From Charles Darwin, *The Expression of the Emotions in Man and Animals* (1872). Reprint, Chicago: University of Chicago Press, 1965. Reprinted by permission of the publisher.

Independent verification that *Vestiges* had significantly influenced Darwin is lacking. Many of the examples previously cited could be found in the natural theology literature, which Darwin did acknowledge in his private notebooks. But the fact that he drew so heavily on the ideas of the natural theologians further illustrates the difficulty in distinguishing science from nonscience. Darwin and the natural theologians used the same kinds of anecdotal animal observations to tell very different stories. For the natural theologian, the complex behavioral adaptations to the environment provided spectacular evidence of God's grand design. For Darwin, such observations were the foundation for a purely naturalistic account of all aspects of behavior, including the development of a moral sense. For Gall and his followers such as Chambers, animal behavior provided evidence in favor of phrenological doctrine. Darwin may have borrowed certain ideas from the natural theology literature, but ultimately his account of animal behavior had far more in common with Gall's naturalistic theory than with that of the natural theologians, even if he refused to acknowledge it publicly.

Darwin recognized that his views would be regarded as materialistic and early on he was already scheming on how to avoid such an attack. In 1838, in the M notebook, he wrote, "To avoid stating how far, I believe in Materialism, say only that emotions, instincts degrees of talent, which are hereditary are so because brain of child resembles parent stock—(& and phrenologists state that brain alters)."[30] Although favorable, such a passing acknowledgment combined with later disparaging references to phrenology indicates that Darwin did not embrace phrenology as it later came to be articulated. Nevertheless, Darwin was clearly indebted to Gall's fundamental idea that the basis of thought and emotion lies in the brain and to Chambers's powerful case for the continuity between animal and human minds. There were good reasons why Darwin would want to distance himself from phrenology. By the time he was writing *The Descent of Man,* with a few notable exceptions such as Alfred Wallace, phrenology had lost essentially all of its reputable support. Darwin's ideas were controversial enough without aligning himself with Gall, a man whose ideas were simultaneously described as atheistic, immoral, and ridiculous.

Just like the ideas promulgated in *Vestiges*, including its phrenological content, evolution was attacked on both scientific and religious grounds. Like phrenology, in the 1820's Lamarckian evolution found its supporters among radicals, dissenters and atheists in London's medical schools and secular universities.[31] Like phrenology, evolution also came to represent both a social and moral philosophy as popularizers, socialists, German intellectuals, spiritualists, and natural theologians distorted Darwin's ideas for their own purposes. In many ways evolution could be regarded as a marginal science—at least initially. However, Darwinian evolution relatively quickly gained scientific respectability. There were powerful interests promoting evolution, and likewise powerful interests also prevented phrenology from moving into the mainstream.[32]

Today, *The Origin of Species* and *The Descent of Man* are still considered classics and outline a research program that continues to be very successful. Except for historians of science, no one reads *Vestiges of the Natural History of Creation* or Gall. Why did phrenology become the example par excellence of a pseudoscience? The answer is not due simply to the "poor" science of phrenological doctrine since the basic ideas of Gall remain valid, but is intertwined with issues of class, the changing meaning of popular science, and the politics of authority and professionalization.

The Rise and Fall of Phrenology

The controversies surrounding phrenology represent a microcosm of the philosophical, religious, political, and scientific debates of the time. For many people phrenology was the key to a much broader philosophy. It went beyond medicine, natural history, or moral philosophy in offering new ideas that would be useful for all classes of society. Gall's comparative approach of examining the faculties and the degree of intelligence of different species gave it more authority than that of previous philosophers who had written on human nature. Gall's theories also provided a way to estimate the motives of human action. By arguing that these qualities were innate, his doctrine provided humans a way to understand the development of genius independent of education or even in spite of an education to the contrary. Gall recognized the philosophical and social implications of his theories, but for the most part, he chose not to promulgate this aspect of his views. That would be left to others, most notably Spurzheim and George Combe. Phrenology eventually took on a life of its own, quite far removed from the researches of Gall.

Spurzheim became increasingly interested in applying phrenology to problems of education, penology, religion, and other nonanatomical concerns. In this new view science and religion merged, and phrenology would reveal God's law of nature. He and Gall began to have more and more disagreements about the nature of the faculties. A fundamental difference between the two men was their view of humankind's essential nature. Gall was not optimistic about the human race, believing that humankind was a sea of mediocrity. Therefore, he emphasized the creative role of genius in the betterment of society and thought the few people who had such extreme talents should be in charge of the rest of the society. He also thought that humans had several evil propensities, including a faculty for murder. Human character resulted from a combination of both good and evil propensities. Spurzheim offered a more optimistic version of Gall's doctrine. He eliminated the faculty "evil" from his system, arguing instead that evil only arose through a mistreatment or neglect of particular faculties, and increased the number of organs from twenty-seven to thirty-five, including adding one called "hope." Furthermore, although people were born with a certain amount of each faculty, he believed they could be modified. Virtues could be cultivated and vices inhibited. Such ideas had tremendous appeal for large numbers of people.

Previous to Spurzheim's arrival in Britain, phrenology had received very little serious attention. Generally Gall was portrayed as a "philosophical charlatan" with his ideas the subject of theatrical farces, satiric poems, and assorted lampoons. Serious articles, when they did appear, were usually highly critical. The *Edinburgh Review* declared: "We look upon the whole doctrine taught by these two modern peripitatics, anatomical, physiological and physiognomical as a piece of thorough quackery from beginning to end."[33] George Combe had been so impressed with the critiques of phrenology that when Spurzheim came to give a series of lectures in Edinburgh, he refused to attend them. However, a friend asked Combe if he would like to come to his house to see Spurzheim dissect a brain. Combe had some training in anatomy and physiology, had often watched brain dissections, but had not been taught anything of brain functions because the lecturer claimed nothing was known. As Combe watched Spurzheim dissect a brain he recognized how "inexpressibly superior" the method was in exposing the detailed structure. He concluded that the press had grossly misrepresented the ideas of Gall and Spurzheim.

Combe decided to attend Spurzheim's second series of lectures, determined to observe and study phrenology himself, and became an avid convert.[34] In 1819 he published *Essays on Phrenology, or an Inquiry into the System of Gall and Spurzheim*, which was subsequently developed into his *System of Phrenology* (1824). In 1820, with some colleagues, he founded The Edinburgh Phrenological Society. In 1828 he published *The Constitution of Man Considered in Relation to External Objects*. Highly influential, it attracted an enormous following, and sold about one thousand copies per year.[35] When an inexpensive "people's edition" became available, in just two years fifty thousand copies were sold. In some households the only other books that were found besides *The Constitution of Man* were the Bible, *Pilgrim's Progress* and Thomas Paine's *Rights of Man*.[36] Within forty years of publication one hundred thousand copies had been sold, more than twice as many as Darwin's *Origin of Species*.

Spurzheim traveled around Great Britain, picking up more and more converts, particularly among medical men. He packed the Hall of the London Institution, drawing the largest audience in the Institution's history. Students in London hospitals paid a "handsome sum" to learn his method of brain dissection. In Cambridge, 130 people attended his lecture demonstrations, including many leading medical men. Previous to 1820 phrenology had been criticized in virtually every medical and literary journal. However, by 1820 many journals, including the prestigious *Lancet,* were asking that Gall's doctrine be critically evaluated. From the 1820s throughout the 1840s phrenology obtained a serious hearing. In the *Lancet* one writer wrote, "It will now be seen whether Phrenology is a 'fraud,' a piece of 'specious quackery,' a system of 'bumps and lumps,' and its pretended discoverers philosophical impostors, or a system founded in nature, and having truth for its basis."[37]

Although Gall's basic physiological premises were often praised, the evidence cited in support of cerebral localization was not. Phrenology's advocates consistently cited the widespread support that the doctrine had among physicians and physiologists. Critics countered that many of these men were not true believers in phrenology's tenets, but merely not actively against it. As scientists they thought phrenology deserved to be empirically investigated, but it would be misleading to claim they were in the phrenologist's camp. The vast majority of articles in support of phrenology consisted of craniological measurements of persons with unusual behavioral characteristics. From the various measurements the strength and weaknesses of the different faculties were determined and then seen if they correlated with the behavioral characteristics observed. However, most examples that were cited as definitive proof of phrenological doctrine were made on classes of individuals whose character and habits in general were perfectly known—convicted criminals and inmates of penitentiaries or madhouses. Phrenology offered no insight into the real character of individuals.

Such criticisms are well illustrated in a case of disease in the organ of combativeness of Mr. N.[38] Mr. N. no longer knew the meaning of words and according to Combe he had become quite irritable, which was confirmed by Dr. MacKintosh who had visited Mr. N on various occasions. On autopsy two lesions were found, a large one in the left corpus striatum where the organ of combativeness resided, and the other in the right hemisphere, supposedly in the area of the language faculty. However, Dr. Craig, the attending physician, had not noticed any change in irritability and nothing could be found in his medical notes referring to Mr. N.'s change in temper. The comments on irritability were from the gardener, and Combe had solicited this information from the gardener's son who was in North America! Dr. Craig pointed out even if Mr. N. had become more irritable, it could be explained by his frustration in not being able to make himself understood and such irritability was expressed toward people he already had reason to be angry with such as the gardener. The irritability had nothing to do with the organ of combativeness. Combe countered that the antiphrenologists had ignored the fact that Mr. N. had knowledge of fourteen languages and that the convolution that corresponded to the faculty of language was unusually large. Mr. N. still knew the meaning of words, but was unable to use them correctly himself. Combe interpreted this to mean that the organ of language had been impaired, but not destroyed. The organ of language was entirely on the right hemisphere; however, the left side contained a cavity with fibers that connected to the right side of the brain. Although phrenology claimed there were separate organs for the various faculties, they were not independent of each other. The major lesion in the organ of combativeness also contributed to the loss of language by way of these connecting fibers to the right hemisphere. Dr. Craig continued to deny that there was any change of temper that accompanied the brain lesion and pointed out that Combe had not made any comments on irritability until autopsy revealed a large lesion that he believed was in the area of combativeness. If phrenology were true, one would expect that the principle seat of the disease would have been in the organ of language. However, the main lesion was on the left side of the brain, in the middle lobe behind the pituitary. The second lesion on the right side of the brain was quite tiny. Combe explained this apparent anomaly by pointing out that the large lesion was directly in line with the fibers that connected to the organ of language and once again placed his emphasis on the area of combativeness.

Mr. N.'s case exemplifies many of the problems that plagued phrenology. First, results were interpreted according to positions already held. Both Dr. Mackintosh and Combe were avid supporters of phrenology while Dr. Craig was highly critical of the doctrine. More serious, evidence for the correlation between character and the various "organs" was often quite poor. As this case illustrated, when contradictions appeared due to additional information revealed upon autopsy, the initial interpretation was modified. However, this is not as damning as it might seem. Scientists continually modify their interpretation of experiments as new evidence comes in. Nevertheless, even in Gall's time the intricacies of brain organization were apparent.

Many scientists, most notably Flourens, vigorously objected to Gall's conclusions. Instead, he argued that the cerebral hemispheres, like other subdivisions of the brain carried out specific functions, but that they were distributed throughout individual structures. Precise localization did not exist. In addition, several other objections were made. First, the formation of the brain did not reveal upon dissections any divisions into organs or compartments. Second, there was no necessary and unvarying relationship between the volume of the brain and intelligence. Third, because of the varying thicknesses of the cranial bones, it was impossible even with the exact instruments of the phrenologists to determine the volume of the brain. Fourth, because of the frontal sinuses and the unpredictable depth of the cranial bones, the development of these various organs, even supposing that they existed could not be ascertained. Finally, destroying or injuring the area of the brain where a particular faculty supposedly was located did not result in the attributes of that faculty being correspondingly impaired.

However, such a seemingly devastating critique does not truly reflect the complexities of the debates over phrenology at the time. First, anyone who studied the brain agreed with the first claim, including Gall and Spurzheim. As Spurzheim admitted, "it is indeed true that the organs are not confined to the surface of the brain. They extend from the surface . . . , and probably include even the commissures, i.e. the whole of the brain constitutes the organs." Yet, if this was the case, one could not expect to accurately measure the "organs" by examining skulls. Phrenologists also would not have disagreed with the second and third points. They recognized that sometimes mistakes could be made in measuring the various faculties because of the problem of the frontal sinuses. Furthermore, the first four objections did not relate to the basic phrenological theories of brain physiology. Thus, many medical people remained quite interested in testing the validity of the theory.

Medical men also dominated the intellectual leadership of a movement promoting the popular dissemination of phrenological ideas to the self-labeled "thinking-classes." Joining forces with practitioners of mesmerism in the 1840s, phreno-mesmerism had a particular appeal because it "offered the first means of giving experimental proofs of the relationship between parts of the brain and particular behaviors."[39] The mesmerist touched the skull at a particular point corresponding to a specific organ and the entranced person exhibited the appropriate sentiment. One subject when the organ of tune was touched she began to hum, time and she tapped her feet, veneration she put her hands together as if she was praying, and touching the skull where destructiveness resided resulted in her pulling and tearing her dress.[40] Doctors became enamored with phrenology in part because they thought it would help maximize their competency to predict and control behavior.

These hopes were not realized. More careful kinds of anatomic studies did not confirm the localization of the various "organs." In 1845 Flourens's experiments on pigeon brains demonstrated that one could remove whole portions of the brain without compromising function. In particular, one could remove the cerebellum (where amativeness resided) without interfering with the power of reproduction. In 1861, Paul Broca with his studies of aphasia showed that the faculty of speech was localized not behind the eyeballs as Gall had claimed, but in an entirely different area. In spite of the detailed claims being wrong, Gall's ideas were the basis for these kinds of studies. He provided a research program that contributed greatly to our understanding of the brain.

Nevertheless, phrenology had an uphill battle in gaining legitimacy, and even at the peak of its popularity always remained on the fringes of scientific respectability with opinions highly polarized. This polarization had more to do with competing class and professional interests than the "facts" of phrenology. Phrenology made various empirical assertions that warranted investigation and it was only later that specific claims were definitely disproved. In phrenology's early days other factors played a more prominent role in its marginalization. In general the "old guard" of the medical establishment did everything they could to discredit it. The majority of physicians who supported phrenology came from relatively deprived social backgrounds, struggling to pay for their medical education, often apprenticing themselves to apothecaries and were working at the lowest ranks of the medical profession. Very few had been educated at Oxford or Cambridge. This was also true of phrenology's other followers. They were often somewhat marginal socially and economically insecure. Few had upper middle-class or aristocratic backgrounds or had elite educations or belonged to elite institutions unlike many of the antiphrenologists. On average the supporters were twenty years younger than their opponents. Most of the antiphrenologists were affiliated with High Church and were supporters of Paleyian natural theology. In contrast, most phrenologists were dissenters and supported a grander view of the Creator. Many were deists. Some were materialists. Powerful social and class interests were working against the acceptance of phrenological ideas.

These different interests were played out in Edinburgh as university professors tried to keep control of what kinds of subjects would be taught at the university and who had authority for investigating the nature of the mind. Thomas Brown, the professor of moral philosophy at Edinburgh University, agreed that the brain was part of the physical world and knowledge of the brain and how it functions was no more atheistic than knowledge of the function of leg muscles. Nevertheless, he was one of phrenology's harshest critics. Generally the supporters of phrenology were "outsiders" who challenged the established norms of science, developing their own standards of scientific explanation based on empiricism and utility. They attacked the moral philosophers who advocated reflection as the primary way to understand the mind as antiscientific, claiming their musings did not rely on observation and had no practical value. Antiphrenologists brought the full weight of the educational establishment to discredit it. Combe was refused permission to lecture at the Edinburgh School of arts and attempts were made to prevent public libraries from purchasing *The Constitution of Man*. In 1836 Combe was a candidate for the chair of logic at Edinburgh University. In spite of the tremendous success of *The Constitution of Man* and over one hundred testimonials in support of his application, he had little support from either the medical establishment or from the fellows of the Royal Society of Edinburgh. Sir William Hamilton, a staunch opponent of phrenology, won the ballot. Phrenology would not be allowed into the hallowed halls of academia.[41]

One way the academic elite tried to protect their interests was to label phrenology a pseudoscience, but the potential of phrenological ideas for understanding the mind was simply too great. By uniting the study of mind with neurology, phrenology cast off abstract metaphysical conceptions of mind. Rather, the mind could be regarded as a concrete organic reality that could be investigated by empirical methods. Countless numbers of societies and journals appeared advocating the application of phrenology to a vast array of social problems. At the peak of its popularity, twenty-nine societies were devoted to phrenology in Great Britain. However, as the interest in phrenology spread to the general public, the medical community became increasingly critical of it. While Gall's basic physiological premises were often praised, his actual theory of cerebral localization and the craniology that went with it eventually was torn to shreds. Although no physiologist denied that a general connection existed between the intelligence and the development of the brain in both humans and the lower animals, "the fixing of the precise localities of their five and thirty 'organs' we regard as one of the bubbles of the day, fit only for grown-up children and would-be philosophers."[42]

The Social Program of George Combe

By the second half of the century phrenology had lost virtually all of its reputable medical support. However, it still had an enormous popular following as numerous journals and societies continued to flourish. Since 1832 *Chambers Edinburgh Journal* had been advocating a program of self-improvement, secular education and the interests of the middle class. Robert Chambers rarely acknowledged the phrenological underpinnings in his articles, fearing the controversy surrounding phrenology would reduce sales and jeopardize reform. But this would change. He founded the *Phrenological Journal*, published an inexpensive edition of *The Constitution of Man*, and in 1836 he wrote a children's book on science for a series of phrenologically based textbooks. He was also secretly planning to write a book on the philosophy of phrenology. That book would eventually metamorphose into *Vestiges of the Natural History of Creation*.[43] However, it was largely due to the untiring proselytizing of George Combe that kept phrenology in the public eye. *The Constitution of Man* argued that the key to a happy life was to live in harmony with natural law. In order to do so one must have an understanding of mind and the science of phrenology would provide that understanding.

Like many of his contemporaries, Combe had become disenchanted with the teachings of orthodox Christianity and was burdened by the doctrine of predestination. Many of the tenets of Christianity were counter to a sense of benevolence and social justice. By the late 1820s he had created a fusion between theological ideas and phrenology, believing that understanding the brain and its function would provide further evidence of God's design. Phrenology could provide the philosophical underpinnings for a true approach to Christianity. Distressed over the many social problems created by rapid industrialization, he saw phrenology as a system that could be used to restore order to what he regarded as a rapidly changing and potentially unstable society. He spent the remainder of his life developing and propagating phrenological doctrine, applying it to a vast range of social and scientific issues.

It is easy to understand phrenology's lure. It represented a popular system of psychology to explain scientifically the great riddle of human nature. In Gall's doctrine there was a denial of original sin or of intrinsic moral failing. Instead, all individuals had the potential to gradually improve themselves as they learned to identify, regulate, and master the temperament with which they each had been endowed. This was the tremendous attraction of phrenology. Phrenologists, like the non-Christian spiritualists, did not think that humans must meekly accept their destiny. Rather, each person could play an active role in molding his or her future, just as humanity collectively could participate in a general moral advance. This was the powerful message of *The Constitution of Man*. Like the plethora of popular psychology self-help books of today, phrenology taught "The Art of Being Happy." It inspired hope for those aware of their deficiencies and was a vehicle of optimism. Humans were capable of modifying their actions. According to Combe, to discover and obey the laws under which nature operated was the key to enjoying life. Although *The Constitution of Man* was filled with references to God and the Creator, Combe did not think that Holy Scripture provided adequate instructions for humankind's time on earth. Rather, God had endowed humankind with faculties to discover the arts, sciences, and natural laws. The physical, moral, and intellectual nature of human beings was open to investigation and once our true nature was understood, this would be our guide for the proper mode of conduct. Many of Combe's ideas were eminently reasonable, and if they had stimulated further scientific research instead of righteous indignation, they could have produced a lot of useful information.

The Constitution of Man advocated a program of reform for virtually all aspects of society. Combe maintained that all institutions, but particularly schools, prisons, and asylums should be redesigned according to observable laws of human nature. Although Robert Owen was quite critical of phrenology because he objected to what seemed an overly biological determinist view of human nature, the reverse was not true. Phrenologists were quite interested in Owen's attempt at social engineering. Combe visited New Lanark, in the early 1820s and was very impressed with what he saw.[44]

Prisons provide a typical example of how societal institutions should be reorganized according to phrenological principles. According to phrenology, a criminal was someone with an unfortunate organization of the brain that could be corrected by applying the principles of phrenology. The conventional theory of penology in the eighteenth century that we see today still (for example, in California's policy of three strikes and you're out) advocated applying severe penalties to crime. Such a policy served as an example to potential wrong doers and would also prevent felons from repeating their crimes for fear of being subjected to the same punishment. This policy of deterrence was based on the premise that the mind of the criminal was the same as that of the average citizen. However, phrenologists believed that the vast majority of criminals had poor impulse control and did not have sufficient moral sense to be inhibited by possible retributions. Thus, harsh punishment would not deter potential criminals and would only make a more hardened criminal of the felon. The more they were compelled and coerced, the worse they would become. Phrenologists, nevertheless, maintained that man's instincts were naturally good; antisocial actions were a form of insanity and should be treated in somewhat the same way. They perceived antisocial activity as a moral problem that should be treated in a "moral hospital." The convict would be considered a "patient" and after a course of treatment the moral faculties would be developed, while unhealthy influences would be removed. The patient could be discharged, not because he had paid his debt to society, but because he had essentially been cured. Prisons should be regarded as rehabilitation centers to give the criminal good treatment, a proper education, and new incentives. Vindictive punishment was not only unworthy of the humanitarianism of an enlightened age, but was useless. Crime was the result of a disordered organization of the brain since the instincts of a healthy brain were only good. Practically, phrenologists differed somewhat as to how these ideas should be implemented, but they were in agreement in the basic scheme. First, the ordinary criminal would be classified phrenologically and given an indeterminate sentence. He would begin by solitary confinement, and gradually be permitted labor and participate in an intensive course in moral, intellectual, and religious instruction. Once repentance and the desire for reformation became apparent, the seclusion and severe discipline could be relaxed. His moral faculties would then be exercised by allowing associations with other convicts of the same classification and at the same stage of reform. As more powers of virtuous conduct became displayed more and more freedom would be granted. Eventually occasional trips away from the prison could be allowed and in time he could be released. One of the virtues of this plan, according to Combe, was that those unfit to return to society would never make the grade. Combe believed that his investigations revealed another class of criminals, what today are called incorrigibles. These were people whose intellect and moral faculties were so small while their more negative propensities were so large that they were irresistibly drawn toward evil. Combe defined them as "morally insane." Ironically, many people objected to Combe's ideas not because they seemed soft, but rather that his reforms went against the reform movement of the time. Advocates of prison reform believed that all men were created equal, which was combined with a religious conviction that salvation was possible for all. Perhaps phrenology was not quite as radical a doctrine as it seemed.

Although the overall message of *The Constitution of Man* was one of optimism and self-improvement, Combe believed there was an upper limit to self-improvement. "Different capacities are bestowed by Nature on different individuals." Since individuals were endowed with different faculties of varying strengths, this meant people were inherently unequal and could be classified into distinct mental types. This mental hierarchy was the basis for a social hierarchy and because one's mental rank could not be changed neither could the social structure be changed. Although phrenology challenged old orthodoxies and vested interest groups, and thus, found wide appeal among the lower rungs of Britain's social hierarchy, its fundamental doctrine of the inherent unequalness of humans meant there would be no alteration of the social structure and no shifts in wealth and power through education. However, there could be an uplifting of the general level of rationality and an associated improvement in morality and social behavior. Rather than revolution, it advocated reform. Phrenology was a "safe" form of radicalism, which also contributed to its widespread appeal.

Unfortunately, another much less savory aspect of phrenology also contributed to its popularity. The inherent inequality of human beings was both gendered and racial. Combe claimed that the Negro was naturally inferior to the white in intellectual capacity because the anterior portion of his skull was somewhat narrower. He did not think this justified slavery—a question that was often asked him when he was visiting America. Instead, he maintained that an extensive program of Negro education was needed and the example of the freed Negroes in the North proved it. But there were more than a few people who were only too happy to accept uncritically such biological arguments of inferiority. Americans were even more interested in what phrenology might have to say about the American Indian who they regarded as a ferocious, "foreign," unassimilable, uncivilized, and unchristianized race. American phrenologists developed huge collections of skulls from various tribes to apply their doctrines and the American journals of phrenology were filled with articles about the American Indian. One of the reasons Spurzheim and Combe visited the United States was to study the psychology of the American Indian. Combe became a friend and avid supporter of the physician scientist Samuel Morton, who collected and measured the cranial capacity of over six hundred skulls. Just like the phrenologists, Morton believed that cranial capacity provided an accurate measure of the brain that had once been inside. He thought it would be possible to rank the races by the characteristics of the brain, particularly size.[45] Spurzheim, Morton, and Combe all concluded that not only was the Indian mentally inferior, but because of the peculiar organization of his mental organs he was intractable and untamable.

The idea of biological determinism has a long and ignominious history. It predates Gall and continues to this day. The tenets of phrenology implied that some people were born criminals due to the particular biological structure of their brains. Women and various racial groups also had a different brain organization than white males and did not measure up. Biological arguments have constantly been used to claim that women and various ethnic groups are innately inferior. The Nazis used biological arguments to justify slaughtering six million Jews along with several other groups of people including gypsies and homosexuals that they deemed inferior. Generally biological arguments have been used to maintain the social status quo, which has systematically oppressed and discriminated against specific groups—at least this is what the standard critiques have been. However, such commentaries miss an important and somewhat paradoxical aspect of why biological explanations have been so attractive. Although Gall was somewhat pessimistic about human nature, Spurzheim and Combe espoused a much more optimistic version of phrenological doctrine. It was the ideas of Spurzheim and Combe that brought phrenology into the drawing rooms of Victorian society, resonating with the Victorians' belief in progress and reform. Phrenology had something for everyone.

The Constitution of Man appealed to upper classes because it reassured them that society's hierarchy was natural and enduring. However, it preached a positive message to other members of society as well. For the middle and professional classes, the meritocratic overtones of the theory meant that one could advance through personal effort, and for the working classes it provided practical hints, aptitude testing, and vocational guidance by which one could improve oneself. The insights of phrenology would allow people to reach their full potential. Therefore, even after it lost most of its following in the scientific community, it continued to have widespread appeal, finding new converts within a progressive reformist framework. This progressive aspect of phrenology became a connecting bond with the whole later spiritualist movement, particularly the non-Christian spiritualists who drew much of their strength from working-class reformist zeal. Phrenology was taught at Mechanic's Institutes and provincial literary and philosophical societies. It allied itself with virtually every reformist group, from antismoking to temperance organizations. While it reinforced racist thinking among those people already so inclined, its overall message was of the innate goodness and the potential for improvement for all members of society. Although the basis of these ideas was rooted in the "science" of phrenology, Combe's program of social reform had very little do with what the growing community of scientists defined as good science. Nevertheless, part of phrenology's widespread appeal to the general public was that it conformed to what many people *thought* was science. Furthermore, both *The Constitution of Man* and *Vestiges* incorporated phrenology within a framework of natural law. This again spoke to the Victorians desire to live a life according to the latest findings in science and at the same time not abandon their faith.

Phrenology's Legacy

As phrenology became more and more popular, it also lost virtually all of its scientific credibility; and the two developments were not unrelated. Late nineteenth-century advocates of phrenology prided themselves in claiming that phrenology was a people's science and would become stagnant if it became incorporated into university study.[46] Although professional and class interests played an important role in keeping phrenology out of the university curriculum, the marginalization of phrenology cannot simply be reduced to a turf fight between philosophers and the naturalists over who should be the authority on theories of mind. Ultimately the "science" of phrenology simply did not measure up. Instead of developing the more difficult kinds of techniques necessary to understand the anatomical underpinnings of specific brain functions, many phrenologists practiced a simple cranioscopy that degenerated into "bumpology" as charlatans offered to measure the bumps on one's head for 25 pence. The claims of phrenology very quickly exceeded the evidence; with many articles appearing that were just plain silly. One article suggested that the faculty for tune was located in the foot rather than the brain, citing as evidence the fact that musicians so frequently keep time by beating the foot.[47] The identification of particular organs and the divisions of the skull can only be characterized as arbitrary. The supposed correspondence between the shape of the skull and that of the brain was quickly used for racist ends. These aspects of phrenology rightly deserve the epithet of pseudoscience. However, focusing only on these elements does not do justice to phrenology's complex history.

Gall pioneered an approach to the study of the mind and brain that was biological, functional, and adaptive. In doing so his work had a profound and lasting influence in three areas. First, the concept of brain localization and the belief that anatomical and physiological characteristics can directly influence mental behavior continue to be the basis for our present-day research on mind and brain. Second, he showed that psychology should be treated as a biological science, rather than as a branch of philosophy. He united all mental phenomena, including the human passions and urged that they be treated consistently in biological terms. In disputing the sensationalism theories of people such as Condillac, he recognized that such theories could not account for the observed differences in talents and propensities between either individuals or species. He also recognized the importance of empirical studies of humans in society as well as species in nature. In his attempt to elucidate the true relationship between the psychological nature of man and that of the lower animals, Gall could be considered the founder of physiological psychology. Finally, the use of his work and its popularization promoted a naturalistic approach to the study of humans that was extremely influential in the development of evolutionary theory, physical anthropology, and sociology.

Phrenology, like Darwinism, represented a line of inquiry that contributed to the separation of psychology from philosophy and the breakdown of the mind/brain distinction. In doing so it added an important voice to the ongoing dialogue concerning the nature of freewill, the soul, instinct, and reason. Just as with the later spiritualist movement, phrenology took on some of the most profound questions that have occupied humankind from ancient times to the present. Phrenology was a science, but it was more than a science. It was both a social and moral philosophy. In it was a plea for humans to learn to act in accordance with moral law. At the same time it was a philosophy of extreme individualism, for the free individual.

The central message of phrenology was that humans could be investigated according to the tenets of science. Like mesmerism, it was a science of possibility. Mental phenomena could be studied objectively and explained by natural causes. Anatomical research could provide a scientific and, therefore, authoritative explanation of human character that could then be applied to many of society's problems. From education to the treatment of the insane, from penology to medicine, phrenology provided a program for human improvement. Like mesmerism, phrenology's tremendous popular appeal lay more in the expectation of what it could offer than being truly a science of the mind.[48] It seemed to provide a panacea, a cure all for all the problems of the common man. However, in this respect, it was not different from many other scientific endeavors. One of the most important legacies from the scientific revolution was that scientific knowledge could be used to improve the lives of humanity. Phrenology offered the hope that such improvement would no longer be limited to material benefits. It might be possible to improve human nature itself.

Gall's essentially biological view of human nature represents the beginning of a research tradition that has been quite powerful and illuminating, and at the same time has led to some very stupid and at times dangerous conclusions. Unfortunately, phrenology's dual legacy can be seen in present-day biological research on behavior wrapping itself in the language of scientific authority of genetics and evolutionary theory. This dual legacy will be explored in the concluding chapter. Phrenology was marginalized while evolution moved to the scientific mainstream, but this has not meant that evolution has been free from ideas that may also eventually be regarded as pseudoscience. Conversely, phrenology's multifaceted history also means that labeling it pseudoscience is also not appropriate.

Combe provides us with a suggestion as to how phrenology might be reevaluated. According to Combe, Gall provided a fresh perspective on the function of the brain, and if Gall's discoveries were true, he believed that history would characterize antiphrenologists as reactionaries impeding scientific progress. The history of science would cast Gall as the Newton of phrenology, and he thought his own role in relation to Gall would be the same as that of Bernoulli to Newton.[49] This did not come to pass. Gall's methodology was crude, and his findings, although detailed, have turned out for the most part to be wrong. Localization of function does exist in a limited sense, but not in the terms that he had used. His errors were conceptual. He claimed that his theories were based on induction, but actually they rested on introspection and on a priori assumptions that the mind could be resolved into a number of independent faculties. Gall did not become the Newton of phrenology as history relegated phrenology to the scrap heap of pseudoscientific theories.[50] But history is continually being reinterpreted, and I would cast Gall in the role of Galileo, rather than Newton. Phrenologists themselves relished such comparisons, adopting the role of martyrs to scientific truth.[51]

Gall, like Galileo, had his work placed on the Index, but there are more compelling reasons that the comparison is apt. None of Galileo's observations were proof of Copernicanism, rather they were merely compatible with a sun-centered system. As definitive proof he cited his theory of the tides claiming that the earth rotating on its axis was what was responsible for the tides. This has turned out to be false; the tides are caused by the gravitational pull of the moon. Gall's basic ideas were correct, but most of his detailed claims have also turned out to be false. Galileo's astronomy was outdated, using the perfect circles of Copernicus rather than the elliptical orbits of Kepler. Likewise, Gall also ignored findings that contradicted his ideas. Rather than trying to develop the physics needed to truly show that the Copernican system was correct, Galileo instead engaged in a battle to convert the Roman Catholic Church to the truth of the Copernican worldview. Gall and his followers also did not do the necessary difficult anatomical and physiological research to put their ideas on a firmer footing, instead devoting their energies to proselytizing the practical benefits of phrenology, urging societal restructuring based on phrenological principles.

Galileo emerges as the key player in the battle over a sun-centered universe because the acceptance of Copernican theory was so intimately enmeshed with issues of religion and authority. It was Galileo's outspoken advocacy of the Copernican system that brought the conflict between the old and new cosmology to a head, challenging the authority of both the Scholastics of the universities as well as the church. This is where the comparison is most powerful. Phrenology challenged the Anglican establishment and the philosophical and scientific views promulgated in the university. Galileo wrote in Italian, not Latin. He wanted his ideas to reach the masses. *The Dialogue Between Two World Systems* (1632) has been referred to as the greatest piece of popular science in the history of science. Both *The Constitution of Man* and *Vestiges of the Natural History of Creation* far out sold *The Origin of Species.* The tremendous impact of these books meant that many of the battles that Darwin had to fight were reduced to mere skirmishes. Phrenology might not have ever been taught in the university, but evolutionary theory also was not incorporated into university curriculum until the end of the nineteenth century. Phrenology may have lost the battle in the halls of the university, but its naturalistic approach would characterize the many new disciplines that eventually became part of a changing university curriculum.

Like the sea serpent controversy, phrenology also demonstrates that the authority of science was highly contested in the Victorian era. As the gentlemen of science were scrambling to gain a monopoly on who determines what is considered legitimate science, both *The Constitution of Man* and *Vestiges* challenged that authority. Part of the uproar over *Vestiges* concerned its anonymous authorship. The emerging scientific establishment tried to discredit it by claiming it was written by an amateur, but no clear-cut professional scientific community existed.[52] Phrenology was the people's science, practical, antielitist, and became closely allied with various reform movements.

Phrenology's practicality and optimism was part of its attraction with its program of self-help and improvement, but its naturalistic underpinnings were problematic, causing major conflicts even among its followers. In spite of the enormous success of *The Constitution of Man*, it also provoked considerable controversy among evangelicals. Natural law as a basis for moral law smacked of determinism and a denial of freewill.[53] By the 1830s the Phrenological Society had lost the support of the Evangelical clergy in the Church of Scotland.[54] Phrenology did not speak to the building crisis in faith permeating all levels of Victorian society. Many Victorians turned to spiritualism to help them cope with what they regarded as an increasingly materialistic world devoid of meaning. The blurring of boundaries between science, theology, marginal science, and pseudoscience is also illustrated in the controversy over spiritualism.

4

The Crisis in Faith

William Crookes and Spiritualism

> Nothing is so difficult to decide as where to draw a line between skepticism and credulity.
>
> —Charles Darwin's perplexed response to William Crookes's investigation of the medium Daniel Home (*More Letters of Charles Darwin*)

The Crisis in Faith

VICTORIANS WERE OPTIMISTIC about the findings of science and at the same time worried that modern scientific discoveries were inevitably leading them to a materialistic worldview. Both the general public and much of the scientific community wanted more than a scientific agnosticism. Spiritualism, especially if it could be proved scientifically, would provide an ideal solution to the fears and anxieties of the time. It assured people that the human spirit survived after death. Psychical researchers in addition to seeing the séance as an opportunity to prove immortality also wanted to explore the enigma of the human mind. Recent developments in both the physical and life sciences lured a few prominent scientists including fellows of the Royal Society, and Nobel Prize winners who supported the Society for Psychical Research into an investigation of spiritualism in spite of it also providing virtually limitless opportunities for hucksters and charlatans.

As was pointed out in the introduction, Victorians found their faith being challenged on a variety of fronts. In theological circles the higher criticism had argued that the Bible was a historical document written by people and should be analyzed as such. Since the Bible was written by different people at different times in history, supposedly reflecting God's will, interpretation of its meaning would always be problematic. Rather than holding onto a rigid unchanging reading of the Bible, its interpretation should reflect the latest scholarship in all disciplines, from history to geology to archeology to astronomy. Darwin's theory provided a completely naturalistic account of speciation, and humans were no exception. If one accepted evolution, then the creation story could not literally be true. Although Darwin was certainly a believer in his early years, he was appalled at many Christian doctrines. In particular, he could not stomach the idea that many of his good friends and relatives with their unorthodox religious views were doomed to eternal damnation. Many Victorians shared his views and for them his theory had a very welcome side to it: Man had risen, not fallen. Instead of being descended from a couple, created in the image of God, and kicked out of Paradise, and still in a state of fallen grace; humankind has gradually worked itself up from miserable animal beginnings to its present status.[1] Thus, some people cast off their faith with a "feeling of exultation and strong hope." Many people also felt that tolerance to a variety of different views was a sign of spiritual maturity.

Even if many people let go of their traditional Christian beliefs, they also found that this created a gap in their lives. If one could no longer accept biblical authority, where was one to turn for guidance on matters of morality and for strength in dealing with the many problems of an increasingly fast-paced and rapidly changing society? Furthermore, while people might be willing to abandon specific Christian doctrine, very few people became atheists or agnostics. Thousands of people turned to spiritualism, which makes it far more representative of religious attitudes reflecting the "crisis in faith" than agnosticism. Although numerous agnostic societies sprung up, primarily intellectuals, and relatively few at that embraced agnosticism. Spiritualism attracted all members of society, including some quite prominent intellectuals. They turned to spiritualism for solace as well as hoping the séance would provide evidence for the existence of the immaterial world, most importantly for the existence of the immortal soul. It provided an alternative to what appeared to be an increasingly bleak, amoral world devoid of meaning. As chapter 3 pointed out, the investigation of the mind from a biological point of view was in its early stages. Psychical research, if not spiritualism, had the potential to contribute to a new science of the mind and also allay the doubts and unease many Victorians felt. Rather than being regarded as pseudoscience, the interest in spiritualism and psychic research reflects the cultural and intellectual concerns of the time and again illustrates the difficulty in drawing sharp lines between science and marginal science. The revival of modern-day spiritualism began not in England, but on the other side of the Atlantic, in a small town in western New York.

The Birth of Modern Day Spiritualism

In 1848 in Hydeville, New York, the family of J. D. Fox had been plagued by mysterious knockings throughout the house. The two daughters, Katie and Margaret, claimed that the knockings were due to some intelligent being, and Katie acted as a "medium" to communicate with the rapper. She was later joined by her younger sister, and the two worked out a code of one rap meaning no, three meaning yes, as well as more complicated sequences. The sisters learned that the rapper was the spirit of a peddler who had been murdered in the house. Later they moved to Rochester, New York, but they continued to communicate with the spirit world. Their rapping sessions began to attract large crowds and in 1851 they went to Buffalo where people paid one dollar per head to hear and see the spirit world unfold before them. In one of the audiences were three physicians from the University of Buffalo who were skeptical of the sisters' demonstration. After a full investigation, they concluded that the knee joints could produce the spirit rappings. Vibrations of nearby articles such as tables and jarring doors could all be produced by the force of the semi dislocation of the bone, which could also control the intensity of the sound. They documented several cases of people who could make such sounds and referred to the sisters as the "Rochester Impostors." The sisters did not take such accusations lightly and asked for an examination to be arranged. However, when the sisters were allowed to sit only in certain positions, the spirits could not be summoned. In spite of a long article in the *Buffalo Medical Journal* detailing how such phenomena could be explained as well as letters to the editor in the popular press, the number of people who believed in spirits continued to grow. The sisters had been exposed as frauds, but they continued to perform to packed halls in Buffalo night after night. Now the spirits no longer just rapped, but they also rang bells and gongs, played the banjo and performed a variety of amazing feats. The sisters no longer allowed investigators into their act and when a man tried to see if the sisters were kicking gongs and cymbals under the tables, he was expelled from the audience.[2]

Certainly, many people regarded the Fox sisters' "show" as just that—entertainment, and did not actually believe that the sisters had summoned spirits from the other world. However, a growing group of spiritualists either deliberately ignored or suppressed the findings on the Fox sisters.

Spiritualism had tapped into a deep-seated need to believe in something—anything. In 1852, Katie Fox surfaced in London as Mrs. Hayden where she attracted a large audience, kindling an interest in spiritualism all over England. In the 1870s, disillusioned, she denounced spiritualism as humbuggery. However, the faithful ignored her, claiming she had become a drunk and couldn't be trusted. The Society for Psychical Research, which attempted to produce reports of high caliber, nevertheless glossed over the fact that Mrs. Hayden had been one of the Fox sisters. In 1888, Margaret also denounced spiritualism as a fraud, explaining how the original sounds had been made from her toe. But such refutations had little impact. Like phrenology, spiritualism had a life of its own, with numerous periodicals and organizations all committed to furthering the spread of the doctrine. More than two hundred groups developed, ranging from provincial societies to London based associations. The vast majority of British spiritualists did not formally join any society, but met informally in their own homes or at local Mechanics' Institutes. Even more people flocked to see the trance speakers who were on the public lecture circuit.

The Phenomena

The core of spiritualism consists of two basic beliefs. The first, quite ancient, is that there is continuity of life after death. Some immaterial essence survives that is referred to as the spirit or soul of that person. Second, it is possible for the living to communicate with these spirits through certain special people—mediums. While certain holy people such as shamans have always claimed that they could make contact with gods and spirits, such people were rare and were usually religious practitioners from traditional cultures. The belief in mediums in Western cultures is quite recent—literally dating from the Fox sisters.

Spiritualist phenomena are classified as either mental or physical. However, since we are dealing with spirits, the division is not entirely clear-cut. Physical phenomena include automatic writing, spirit photography, acoustic phenomena, glowing lights, apports (objects that were not in the room, but suddenly appear in a séance), levitation of people or objects, telekinesis (movement of objects by inexplicable means or the ability to cause such movement), ectoplasm productions, stigmata (marks or sores that represent and resemble the crucifixion wounds of Jesus), trances, and elongation of the body. Mental phenomena include telekinesis, clairvoyance (the power to perceive things that are out of the range of the normal human senses), and divination (the act of foretelling future events or revealing occult knowledge).[3]

Most relevant to the development of mainstream spiritualism were the séance mediums. These people sat around a table holding hands with other sitters. The lights were low and often music was played. The medium would then go into a "trance" and would speak, giving information or an impersonation of the dead relatives of the sitters. Sometimes "spirit" hands would touch the sitters, apports might appear, and these objects would usually have some significance for either the sitters or their dead relatives. Some mediums produced spirit art or specialized in slate writing. The most dramatic of all the séance phenomena were materializations. "From the bodies of some mediums a strange foam, frothy or filmy substance, dubbed ectoplasm might be seen to condense."[4] If a sitter was extremely fortunate, she might witness a partial and then a full materialization of a dead relative. The procedure for these materializations followed a standard pattern. The medium would be tied down in a cabinet, and a few minutes later a fully materialized spirit of a deceased person would appear. After the spirit left a few minutes would pass, the cabinet would be opened, and inside was the medium tied up, just as before. However, many "spirits" were grabbed by rude sitters and were discovered to be the medium in disguise. Although the vast majority of mediums were exposed as frauds, a few became quite famous including Douglas Daniel Home and Florence Cook, who seemed capable of bringing forth truly astounding phenomena.

The chemist and physicist William Crookes became interested in spiritualism for both personal and scientific reasons, curious to find out if any of the phenomena were legitimate. He investigated many different mediums, but his most detailed experiments involved Home and Cook. His investigations of Cook jeopardized not only his scientific reputation, but also cast doubts on his own personal conduct. He actively researched psychic phenomena for only a short period of time, from 1870 to 1875, but the controversy surrounding his involvement with spiritualism provoked debate among spiritualists, antispiritualists, scientists, and historians of science. In trying to unravel the many strands of Crookes's interest in spiritualism, the difficulties in drawing boundaries between science and nonscience become readily apparent.

William Crookes, Florence Cook, and D. D. Home

William Crookes was born in 1832, one of sixteen children. He had somewhat irregular schooling, but eventually became a student of A. W. Hoffmann at the Royal College of Chemistry, where he received a rigorous training in analytical techniques. He became Hoffmann's personal assistant and came to the attention of Michael Faraday. Crookes modeled himself after Faraday and like Faraday was an excellent lecturer and an outstanding experimentalist. He was elected to the Royal Society in 1863 for discovering the element thallium two years earlier. In investigating the properties of the new element, particularly determining its atomic weight, he set extremely high standards of precision for Britain's foremost chemical analysts. Although trained as a chemist, Crookes also achieved remarkable stature as a physicist. He pioneered experiments that laid the groundwork for J. J. Thomson's discovery of the electron for which he was awarded the Nobel Prize. Indeed, Thomson claimed his own researches were inspired by Crookes's "beautiful experiments on cathode rays."[5] Crookes's scientific credentials were impeccable. This was no "marginal" scientist, but he was drawn to spiritualism knowing full well that it was a haven for charlatans and frauds.

Séances had become quite fashionable in Victorian society by the late 1860s. Crookes appears to have been initially attracted to spiritualism when his youngest brother, whom he was quite close to, died of yellow fever. Brought up with the traditional Christian belief in the afterlife, Crookes was persuaded to attend a séance in 1867 to try to make contact with his brother. He became quite interested in the kinetic, audible, and luminous phenomena that he witnessed at séances. In 1870, he publicly announced that he planned to make a serious investigation of spiritualism, which he described as "the movement of material substances, and the productions of sounds resembling electric discharge." He, however, maintained, "I have seen nothing to convince me of the truth of the 'spiritual' theory." In a paper contributed to the *Quarterly Journal of Science* (July 1870), he wrote of the rapping sounds often heard at a séance:

> The scientific experimenter is entitled to ask that these taps shall be produced on the stretch membrane of his phonautograph.
>
> The spiritualist tells of heavy articles of furniture moving from one room to another without human agency. But the man of science has made instruments which will divide an inch into a million parts; and he is justified in doubting the accuracy of the former observations, if the same force is powerless to move the index of his instruments one poor degree.
>
> The Spiritualist tells of flowers with the fresh dew on them, of fruit, and living objects being carried through closed windows, and even solid brick-walls. The scientific investigator naturally asks that an additional weight (if it be only the 1000th part of a grain) be deposited on one pan of his balance when the case is locked.[6]

Crookes wanted to bring the methods of scientific investigation to the field of spiritualist inquiry in search of the truth and to "drive the worthless residuum of spiritualism hence into the unknown limbo of magic and necromancy."[7] This was his position for the next five years, as he investigated various mediums. Skeptical and quite appalled at the many fraudulent practices, he, nevertheless, became convinced that a few mediums, most notably the Scottish American Douglas Daniel Home, were genuine. In a room he had built onto his house specifically for his spiritualist investigations, Crookes had a series of séances with Home between 1871 and 1873.

Home was the most famous of all the nineteenth-century mediums, provoking the most commentary, not only in the spiritualist press, but in the leading Victorian periodicals as well. Home possessed a 'psychic force' that could produce musical effects, modify gravity, and various other phenomena that were unknown to either magicians or scientists. He had enormous talents, and no medium had more devoted followers. His séances were free, but he accepted extravagant gifts from wealthy patrons. In 1868, he was charged with trying to defraud a rich old widow of thousands of pounds. *Punch* quipped that "Spirit hands at his bidding, will come, touch, and go/But you mustn't peep under the table you know." Nevertheless, no one was ever able to prove unequivocally that Home was a charlatan and to this day he remains an enigma.

Home collected the rich, the famous, and the powerful wherever he went, being equally at ease in English drawing rooms, Italian villas, and royal palaces. Throughout the 1860s, he conducted séances in the homes of numerous English aristocrats, for many famous literary figures, and cooperated when scientists wanted to investigate his powers. He was also accessible to the common folk as he held sittings for free.

In many ways, Home's séances were not all that different from numerous other mediums. However, a couple of practices he did later on in his career put him in a class of his own. One was body elongation. Several witnesses claimed that his body grew in length, anywhere from several inches to nearly a foot. He was also able to handle red-hot coals drawn from the fire with no apparent discomfort or after effects. His most famous event was the Ashley House levitation, which supposedly occurred in December 1868. According to witnesses, in a trance Home floated out the window of one room overlooking Victoria Street and entered through the window of the adjacent room, where startled witnesses were assembled. Is there another interpretation of these phenomena? Was Crookes's investigation of Home capable of detecting "tricks?"

Crookes's sittings with Home were in many ways typical of his approach at other séances. Although it appeared that Crookes had clear scientific control over the proceedings, on close examination this was not really the case. He was confident that his use of mechanical devices prevented any chance of deception, but this was not true. Crookes always tried to have trained observers present whenever possible, but Home did not allow him and his scientific friends any real authority over the conditions. For instance, if the spirit commanded "all hands off the table," all observers, including Crookes, removed their hands. Home often imposed his own conditions while seemingly cooperating with the scientists. If he didn't like Crookes's arrangements, and no phenomena occurred, Home would just say that circumstances were not congenial. Thus, it is worth examining a few of the more famous phenomena that Home demonstrated and see how they might be explained.

Several reports exist of an accordion (actually a concertina) being played while held by only one hand on the non-key side or when no one was holding it at all. Common sense suggests that it is necessary to have two hands to move the bellows in and out. Crookes placed one of Home's hands in a wire cage while Home held the accordion vertically on the non-key side, but it still played. After Home's death, a small harmonica that played one octave was found among his personal effects. The harmonica sounds quite similar to a concertina and could easily have been concealed in Home's mouth. People had an expectation of hearing the accordion, Home had a bushy mustache, and in the dark it would be easy to fool people. The movement of the bellows that was sometimes seen could have been managed by "catching" the lower end of the instrument upon some hook or protrusion and thus, a simple wrist action could produce the movement.

Many eyewitnesses testified to seeing Home elongate his body as much as eleven to twelve inches on several different occasions. An explanation for this phenomenon was revealed in the vaudeville acts of the twentieth century. One of the most famous acts was by a man named Clarence Willard who was known as "The Man Who Grows." His elongation was not fake, although his performances were aided somewhat by an optical illusion. Willard explained his technique to a meeting of the Society of Magicians in 1958, and he claimed he had added two inches to his height by constant stretching. It is possible for the top of the hipbones and the short ribs to separate. In Home, they were unusually close together, which meant there was more potential for stretching and it is quite likely that he used a similar technique to the one that Willard used decades later.

On numerous occasions "spirits hands" appeared, which were usually glowing and often touched the various people around the séance table. Home's hands were visible on the top of the table and not being held by anyone. The spirit hands often ended at the wrists. Nothing else could be seen and the hands often seemed to melt away when people attempted to grasp them. As was customarily the case when unusual phenomena occurred, the lights were low. Crookes described what he had seen, which was accurate enough. However, he assumed what he was seeing was genuine and did not seriously try to find out how he could have been deceived. Home was an excellent sculptor and had made his living as one for a short time in Rome. A person who had visited Home's studio in Rome claimed it was filled with sculpted hands. It would have been possible for

Home to substitute wax hands on the top of the table, leaving his real hands free to do whatever they wanted. Mr. F. Merrifeld reported in the *Journal of the Society for Psychical Research* that at one of Home's séances "the 'spirit hand' rose so high that we saw the whole connection between the medium's shoulder and arm and the 'spirit hand' dressed out on the end of his own."[8] Home was never searched before or after a séance. The glowing or light-emitting hands could be explained by rubbing oil of phosphorus on his hands. We know that Home had experimented with oil of phosphorus. Coating an object with calcium or zinc oxide and then exposing it to a bright light will also cause the object to emit a light for a short period in the dark and then slowly fade out. This fading out was exactly the phenomenon that was observed concerning the spirit hands on several occasions.

As previously mentioned, the most famous of all of Home's manifestations was the Ashley House levitation. Once again, a close examination reveals all kinds of discrepancies bringing into question what the eyewitnesses had actually observed. First, a "dress rehearsal" occurred in November. Home was at the window that he opened and he went outside on the ledge. A spirit talking through Home then explained how Home might be lifted in the air by first being on the back of Lord Adare's chair. Adare felt his feet there and then Home appeared to be lifted up and carried to the other side of the room. This was done in complete darkness, not only making interpretation of what actually happened difficult, but also undoubtedly priming the sitters for the main event coming up in a few weeks.

According to Adare, Home was in an elongated state, pacing around the room. He left and the sitters heard him go into the other room and open the window. He then appeared outside the window of their room and climbed in. When they went to the next room, the window was only slightly open. Home had told Adare to go to the other end of the room and he then levitated and floated out horizontally through the narrow opening of the window. However, the testimonies of the three witnesses reveal numerous contradictions concerning the date, the location, the brightness of the moon, and the height of the windows from the ground, the size of them, and how far apart they were from each other. Most serious are differences in the actual descriptions of the event. Alternative explanations of what might have happened are as follows. There were steel bolts protruding on either side of the balconies and it might have been possible for Home to fasten a heavy cord between the two bolts that would have allowed him to edge his way from one balcony to another. He also could have opened the window in the other room, but then sneaked back into the first room in the dark. He then stepped up on the windowsill inside the window and opened the window in back of himself and simply stepped into the room from the windowsill. In the dark it would have been hard to discern whether Home was actually on the outside or inside of the windowsill. These explanations do not account for Adare having seen Home levitate horizontally in and out of the window in the other room. However, no one actually saw Home go out the window of the other room the first time. We only have Lord Adare's account that Home showed him what he had done, and Adare was a close personal friend of Home.[9]

The evidence for Home's other levitations consist of his word that he had levitated, along with witnesses claiming that his boots could be felt suspended in the air. In a darkened room, a medium could remove his boots, climb on a chair and hold them up and allow people to feel the boots while he holds the top of them. Standing on the chair, he could project his voice. With a piece of chalk on a rod, he could mark the ceiling, which was the "proof" that he floated to the top of the ceiling.

Although nothing was ever definitively proved against Home, many people were skeptical of these accounts, even at the time. In addition to Mr. Merrifeld, a particularly suspicious incident involved a Mrs. Lyon. A wealthy and credulous widow, Mrs. Lyon received several messages from her husband through Home to transfer £24,000 and then a further £30,000 to Home as well as smaller amounts. She successfully brought suit against Home that it was the medium himself who had given the messages and he was ordered to make a restoration of £60,000. Although the case was decided in her favor, this was not definitive proof that the messages did actually emanate from Home. The suit certainly damaged his reputation, but the Ashley house levitation more than counteracted the negative publicity from it.[10] Furthermore, Mr. Merrifeld did not publish his accounts until 1903. At a time when fraudulent mediums were being exposed right and left, no one was ever able to catch Home in an act of deception.

Home captured the minds and emotions of audiences for over twenty years, but the circumstances surrounding him remain a mystery. While a variety of explanations have been put forth, none have been entirely satisfactory. Humans have an enormous potential for self-deception and it is quite possible that a skilled medium such as Home could bring about something like group hypnosis at a séance. Apparently, Home saved his special effects for a select few of his admirers. These were people who could be psychologically primed to anticipate miracles and, thus, be convinced that they had indeed witnessed them. In addition, there were undoubtedly servants who could be bribed to help prepare hidden mechanical devices in the séance room. A truly bizarre theory was put forth by the anthropologist E. B. Tylor, who tentatively suggested that Home might be a werewolf with "the power of acting on the minds of sensitive viewers."[11] Clearly, Home was an exceptional person and he had the power to act on the minds of susceptible spectators. Whether Home's power came from what we today call mind control, or from religious and inspirational messages that regularly marked his séances, or the willingness of his clients to believe, or as his supporters claimed from the authenticity of the phenomena associated with him, we do not have a complete explanation. Followers and critics of Home, of course, have different theories about his powers. But it is true that nothing succeeds like success. The vast majority of mediums had been exposed as charlatans. Yet, Home was never publicly exposed as an impostor. In fact, he was very critical of many mediums.

In 1877, Home published *Lights and Shadows of Spiritualism*, which gave a history of spiritualism and extolled the evidence for genuine spiritualism. However, he also devoted three chapters to exposing the tricks of mediums. He had harsh words in particular for those mediums manifesting full-body materializations, which he described as "the form of fraud at present most in vogue."[12] Home's willingness to disclose these secrets angered many mediums, while at the same time convinced many people that his phenomena must be genuine.

In addition to Home, Crookes had a series of séances with the medium Florence Cook between 1874 and 1875. A swirl of controversy surrounds Crookes's investigation of Cook. Her reputation was not nearly as untainted as Home's and threatened to severely damage Crookes's reputation as well. Indeed, Florence Cook illustrates the seamier side of spiritualism. Cook was a young teenage medium from Hackney who specialized in full-body materializations. She caused considerable consternation among spiritualists, psychical researchers, and their critics.

Cook's embodied spirit was "Katie King," who quickly became a familiar name in Victorian spiritualist households. Home always remained among his sitters (except in the Ashley house levitation), but Cook's phenomena required that she withdraw from the séance table and enter a cabinet, cupboard, or portion of the room that was partitioned off with a curtain. She was seated on a chair, tied up with a string whose knots were sealed, and remained there for varying lengths of time while the sitters would compose themselves, often with spiritualist hymns. If conditions were favorable, Katie's face would emerge at an opening in the curtains or at the top of the cabinet. Sometimes several different faces would appear alternating with Katie's. At other times accompanying spirit hands would accept paper and pencils from the sitters and scribble messages. Most impressive, however, were the occasions when a fully materialized Katie would appear and mingle among the sitters. Crookes investigated these full-body materializations and seemed convinced that they were genuine, claiming that he was once privileged to walk arm in arm with the attractive spirit (see figure 4.1).

The opportunities for deception were enormous with full-body materializations—so much so that some mediums resented the ill repute that the shady phenomena brought to the entire profession. Sergeant Cox, a committed spiritualist, wrote Home a long letter describing how such a

FIGURE 4.1. Exposure of a Medium. The sequence of events depicts Sir George Sitwell's exposure in 1879 of a medium masquerading as a materialized spirit and was illustrated in the *Graphic*. Published in Janet Oppenheim, *The Other World: Spiritualism and Psychical Research in England, 1850–1914* (Cambridge: Cambridge University Press, 1985). Reprinted by permission of the publisher.

hoax could be achieved, which Home published in *Lights and Shadows of Spiritualism*. Cox had gotten the description from a medium, who had written another medium how to pull off the illusion.

> All the conditions imposed are as if carefully designed to favour fraud. . . . The curtain is guarded at either end by some friend. The light is so dim that the features cannot be distinctly seen. A veil thrown over the body from head to foot is put on and off in a moment and gives the necessary aspect of spirituality. A white band round head and chin at once conceals the hair, and disguises the face. A considerable interval precedes the appearance—just such as would be necessary for the preparations. A like interval succeeds the retirement of the form before the cabinet is permitted to be opened for inspection. This just enables the ordinary dress to be restored. While the preparation is going on behind the curtain the company are always vehemently exhorted to sing. This would conveniently conceal any sounds of motion in the act of preparation. The spectators are made to promise not to peep behind the curtain, and not to grasp the form. They are solemnly told that if they were to seize the spirit they would kill the medium. This is an obvious contrivance to deter the onlookers from doing anything that might cause detection. It is not true. Several spirits have been grasped, and no medium has died of it; although in each case the supposed spirit was found to be the medium.[13]

Cox then described how the medium concealed a thin veil in her drawers. She wore two shifts under her gown. She took off the gown and carefully spread it on the sofa over the pillow so as to look like the medium while the real medium emerged in the shift with the veil. Sometimes a person was allowed to go behind the curtain when the "spirit" was out in front and attested that they saw or felt the medium. However, Cox reluctantly concluded that virtually always the favored person was either part of the deception or was a close friend and undoubtedly would not be willing to expose the friend as a cheat. He concluded that no phenomena should be accepted as genuine that are not produced under strict test conditions. The question could be unequivocally decided simply by drawing back the curtain while the alleged spirit is outside and showing the medium inside to the eyes of all present. Cox offered a reasonable solution to the controversy surrounding materializations, but one that no one seemed to be willing put into practice, including Crookes.

Crookes, this most eminent of scientists who was famous for his powers of observation and analysis, alleged that on several occasions he saw the fully materialized form of Katie King together with the medium Florence

Cook. He reported in the *Spiritualist Newspaper* that he clasped "Katie and felt a substantial body," but he also maintained that she was "pure spirit." He asserted that he had repeatedly photographed Katie.[14] What are we to make of such claims? Most people think either the man was utterly gullible to believe in such crass deception or he was part and parcel of the fraud, playing a huge joke on the public, and this was perhaps due to a romantic entanglement with Cook. It has been suggested by Trevor Hall that Cook was William Crookes's mistress and that he was her confidant in several blatantly fraudulent séances.[15] Florence was certainly an attractive young woman, and perhaps Crookes was unable to resist an opportunity to flirt with her or Katie during the séances. Such a position outraged not only some spiritualists, but also particularly upset people who held Crookes's name in high regard. We do not know for a fact that Cook was Crookes's mistress and will examine the evidence for such a claim in a moment.

We do have evidence, however, that Florence was a fraud and had been caught in several deceptions. In 1873, William Volckman gained admittance to a séance with great difficulty. Katie had been present for about forty minutes and Volckman became convinced that she was Florence and he grabbed the "spirit." In the struggle that followed a very physical Katie got away with the help of several of the gentleman sitters, including Edward Corner who would marry Florence four months later. Katie rushed back into the cabinet. About five minutes later when the cabinet was opened, a very disheveled Florence was found, still tied with the same tape with which she had originally been bound.[16] On another occasion Katie paraded around in Crookes's laboratory arm in arm with Mary Showers's materialization, named Florence Maple. Not only had Mary Showers been caught in many fraudulent séances, but Sergeant Cox also wrote a devastating account of this incident:

> They breathed, and perspired, and ate. . . . Not merely did they resemble their respective mediums, they were facsimiles of them—alike in face, hair, complexion, teeth, eyes, hands, and movements of the body . . . ; no person would have doubted for a moment that the two girls who had been placed behind the curtain were now standing in *proriâ personâ* before the curtain playing very prettily the character of ghost.[17]

Furthermore, as Cox pointed out, all that would have been necessary to definitively prove that the "spirits" were genuine was show that Miss Cook and Miss Showers were asleep on the sofa behind the curtain, but such evidence was not offered or even allowed.

Given these accounts, what kinds of evidence did Crookes have indicating that the materializations were genuine? Crookes attached Florence to a galvanometer (which measures electrical currents), and he assumed that she could not impersonate a spirit without producing telltale fluctuations in it. Katie, nevertheless, emerged. He attached the galvanometer to another medium, Annie Fay, as well. Spirit hands seemed to emerge, the spirit threw a box of cigarettes, and musical instruments all played at once, but the galvanometer still held steady. Crookes believed he could prevent all forms of possible fraud by using these electrical tests. However, later psychical researchers pointed out that both Cook and Fay could have used other parts of their bodies or even a resistance coil to maintain the electrical current intact, which would then free their hands for other purposes. While this does not explain the materialization of Katie, it does suggest that Crookes had less control over the experiment than he confidently assumed.

Crookes was an excellent photographer and claimed that he had taken forty-four spirit pictures, but most of them had been accidentally destroyed, and he did not allow the remaining ones to be publicly viewed or published. These photos showed Katie King and another person either lying down or in a chair, supposedly Florence in a trance. However, no photos showed the faces of the two women at the same time. Katie was either in front of Florence's head blocking it or Florence's head was covered by a cloth. Katie also varied in size and shape on different occasions and thus, it seems that there were different Katies.

It appears that "Katie" was either the medium herself or at other times a partner in cahoots with Florence. The latter seems most likely in the many séances that Cook held in her family's house. Katie's appearance was described quite differently on a variety of occasions. However, in Crookes's own home such a hoax could not have been perpetuated without his consent. Thus, it is more likely that Katie was Florence in disguise. Nevertheless, Crookes declared, "I have the most absolute certainty that Miss Cook and Katie are two separate individuals so far as their bodies are concerned."[18] Here was a man with a flawless scientific reputation, who discovered a new element, but could not detect a real live maiden who was masquerading as a ghost.

The claim that Florence was Crookes's mistress has not received much serious attention, the consensus being that he was not a knowing partner in Cook's scheme, but that he was duped. He certainly was neither the first nor the last scientist whose self-confidence and technical expertise gave him a false sense of security and led to self-deception. He did not think he had to be fully accountable to his readers as he did not leave any records of his séances with Florence. He somewhat arrogantly admitted that he was "accustomed to having his word believed without witnesses" and grew quite annoyed when questioned whether he could have possibly been mistaken about the events he witnessed. Quite fond of Florence, he had absolute faith in her honesty. Katie's confidence grew in Crookes and he wrote in the *Spiritualist Newspaper* that she wouldn't give a séance unless he took charge of the arrangements. It appears, however, that it was Katie/Florence who was actually in charge. She specified certain aspects of the séance, such as who could sit where and what kinds of investigations into the phenomena would be allowed. Crookes in many ways was perhaps an ideal candidate to be tricked by Florence. His excellent reputation and his vanity made him think he could not be deceived. This allowed him to support her powers of mediumship unequivocally and gave a tremendous boost to her career.

Evaluating the evidence concerning Crookes's relationship with Florence illustrates many of the factors that come into play and the difficulties in determining the legitimacy or illegitimacy of scientific practice. There are three possibilities in regard to Crookes's investigation of Cook's mediumship. First, the phenomena as reported by Crookes were genuine. Second, they were fraudulent, but that Crookes was completely deceived by Florence and believed that they were genuine. Third, not only was Florence a fraud, but Crookes also aided and abetted her deception by his reports.[19]

Crookes had a serious interest in investigating spiritualist phenomena in a rigorous scientific manner. Furthermore, he was skeptical of the phenomena as clearly documented by his article in the *Quarterly Journal of Science*. It is also equally clear that such high standards of investigation were not maintained. Were the séances genuine or not? From a variety of different kinds of evidence, it is virtually impossible to conclude anything but that Florence was a fraud. Her sister and her friend Mary Showers had been caught in numerous deceptions. She was trained by and associated with mediums known to be tricksters. A variety of accounts in spiritualist publications suggested that she had engaged in palpable fraud. In addition, Crookes's own reports of the séances in the *Spiritualist Newspaper* were quite problematic. Crookes had written that he alone was privileged to enter the cabinet as he pleased during the sittings. At one sitting he had assisted Katie by lifting the other form into a position of safety on the sofa. This other form was either a confederate or a bundle of clothes covered by Florence's dress. Since this sitting was at Crookes's own home, it is difficult to imagine how a confederate could have been sneaked into a locked room to which only he had the key without his knowledge. If there was no other person involved, then the reclining figure was a bundle of clothes. Once again, it is hard to believe that Crookes would not have been able to tell the difference between a bundle of clothes and a live body. In the case of Katie King and Florence Maple walking arm in arm, no other witnesses were cited to offer corroborative evidence. Crookes's own lax accounting of the event combined with Cox's description of the Cook/Shower séance make it difficult to believe that Crookes could have been an innocent dupe. How could Crookes have allowed himself to become caught in such a deception? Why would he completely abandon the admirable and critical standards of evidence in psychical research that he had described in the *Quarterly Journal of Science* article? According to Hall, only a sexual entanglement could explain so fantastic and undignified a transformation.

In the mid-1870s three scientists all made statements in print that Crookes was hopelessly infatuated with Florence. Crookes's own description of Katie in her final appearance seems to suggest this.

> But photography is as inadequate to depict the perfect beauty of Katie's face as words are powerless to describe her charms of manner. Photography may, indeed give a map of her countenance; but how can it reproduce the brilliant purity of her complexion . . .

As a eulogy for Katie he cited a poem:

> Round her she made an atmosphere of life,
> The very air seemed lighter from her eyes,
> They were so soft and beautiful, and rife
> With all we can imagine of the skies;
> Her overpowering presence made you feel
> It would not be idolatry to kneel.[20]

Was such an obviously smitten Crookes capable of objective investigation? Documentation exists that Florence accompanied Crookes to Paris on more than one occasion. The final type of evidence in support of a liaison between Crookes and Florence were statements by two men who had been lovers of Florence at the end of the nineteenth century. They both testified that she had told them that she had been Crookes's mistress at the time of the 1874 sessions and that they had used the sessions to conceal their liaison. Certainly, there is a possibility that Florence lied to these men. Her reputation for honesty was less than Sterling. The possibility also exists that the two men's statements were not entirely accurate. Finally, one could argue that if Crookes and Florence really were lovers, this makes it more, rather than less likely, that he was deceived and not part of the fraud himself. Crookes's investigations of Mary Showers made him realize that she consistently produced fraudulent manifestations. This was extremely upsetting to him, as she and Florence were friends. Although Crookes never specifically repudiated Cook, he did suggest that his faith in her genuineness was shaken. He certainly never would have acknowledged that he and she had been lovers. He also would not want to admit that he could have been duped by a clever medium. Even if the actual relationship between Crooke and Cook could be ascertained, it would raise as many questions as it would answer. Both committed spiritualists and scientists have been quite angered by such a suggestion. Scientists are reluctant to admit that one of their own was guilty of scientific fraud in order to perpetuate an adulterous affair. However, if this was the case and Florence was a trickster, it would mean that the phenomena were not genuine. Such a conclusion is unsatisfactory to the spiritualist community as well.

At the beginning of his investigations, Crookes wrote that spiritualism "more than any other subject lends itself to trickery and deception," and he never had any reason to change that assessment. By the end of his active involvement with spiritualism, he had investigated over one hundred different mediums, and only Home remained free of suspicion. In a letter to Home in 1875, he complained, "I am so disgusted with the whole thing that, were it not for the regard we bear to you, I would cut the whole Spiritual connexion, and never read, speak, or think of the subject again."[21] Only a few months after his published eulogy of Katie, he refused to discuss the Katie King materializations.

Obviously disillusioned, Crookes wanted to move on. The mainstream scientific community may have regarded his spiritualist investigations with dismay, but they also regarded his other work highly. He was knighted in 1897, received the Order of Merit in 1910, and was president of the Royal Society from 1913 to 1915. Crookes's investigations of spiritualism cannot simply be regarded as a passing fancy in which his critical judgment temporarily went awry because of his infatuation with a young beautiful medium. His beliefs concerning different phenomena seemed to vary over time with the emotional circumstances of his personal life. Although he was initially drawn to spiritualism after the death of his younger brother, by 1874 he had abandoned the spiritualist hypothesis. Nevertheless, he continued to be interested in psychic phenomena. At the end of his life when his wife died, he seemed to revert to pure spiritualism. Perhaps intense bereavement loosened some restraints he usually had as a researcher, which in turn made him susceptible to Florence's wiles. However, a variety of other phenomena attracted Crookes's attention that he continued to investigate throughout his life.

While Crookes may have concluded that spirit manifestations were all fraudulent, this did not rule out the existence of some undiscovered psychic force. In many of his various experiments on Home, he was trying to discern whether Home exuded such a force. He placed the medium in a helix of insulated wire through which he passed electric currents of different intensities. He brought strong magnets near Home and to objects that had moved in Home's presence. He had illuminated experiments with different colored lights to see if that had any effect, but he was unable to identify anything that might be responsible for Home's abilities. Nevertheless, he never gave up believing in a psychic force. In 1887, he became president of the Society for Psychical Research. By the end of the 1880s, he had come to the conclusion that the more we knew only revealed how much more we didn't know. As he was preparing his upcoming address as president of the British Association for the Advancement of Science, he admitted that he wanted to discuss occult phenomena, but he also realized that he didn't have anything definitive to build his remarks on. Although he never ceased believing that psychical research would eventually reveal the operations of a new force, he became less and less certain that this force would find its place in modern physics.

Crookes turned his attention to the study of the mind. Rather than trying to discover a physical force, he thought that a deep understanding of the brain would provide the connecting link between mind and matter. In his address to the BAAS, he admitted that if he was beginning his investigations again, he would put much greater emphasis on trying to understand telepathy. He believed that it was a genuine phenomenon in which thoughts and images could be transferred from one mind to another by some as yet undiscovered force or sense. He thought that telepathy operated according to some fundamental law, just like gravity or electromagnetism. Perhaps the recently discovered roentgen, or X-rays, might be involved in the transmission of telepathic messages. Although his ideas were highly speculative, he also thought it was unscientific to call in the aid of mysterious agencies. Crookes was trying to bring psychic phenomena into the realm of physical law. However, at a meeting of the Ghost Club in 1899, he also claimed that spiritualist phenomena did not appear to operate according to natural law, that is, certain causes are followed by certain effects. Thus, while he explored the possibility of links between roentgen rays and the cerebral ganglia, he also speculated that spiritual beings were "centers of intellect, will, energy, and power, each mutually penetrable, whilst at the same time permeating what we call space." These centers each preserved their "own individuality, persistence of self, and memory."[22] Crookes may have dismissed as inadequate the evidence that genuine communication with the dead had occurred in séances, but this did not rule out the possibility for him that some sort of spirit entity survived after death.

Crookes is certainly one of the more fascinating characters in the history of science, and his involvement with spiritualism dramatically illustrates that psychical research tests the limits of skepticism as well as credulity. It is not possible to make a sharp division between Crookes's physical and psychic experiments. Both were motivated by a curiosity and a desire to explain the forces that shaped the universe. Although the vast majority of spiritualist phenomena were indeed fraudulent, nevertheless, Crookes along with other physicists thought there were aspects that warranted serious scientific investigation, and again illustrate the difficulties in drawing sharp boundaries between "ortho" and "pseudo" science.

Physicists and Psychic Phenomena

Crookes was not the only physicist who believed that investigating spiritual and psychic phenomena was the next logical step in extending our knowledge of the universe. The nineteenth century produced a vast increase in our understanding about what we know about matter. Much of that understanding dealt with the discovery of aspects of matter that had been quite mysterious: the nature of energy, heat, magnetism, electricity, light, and the relationship between these various phenomena. Physicists had described how these entities were related to one another in the form of laws and models, which further reinforced the idea that the natural world was orderly and knowable. The discovery of radioactivity, X-rays and electrons also suggested the possibility of still other undiscovered new forces. These findings resulted in physics undergoing a profound revolution, shattering the certainty implied in the Newtonian worldview. Instead of a universe that operated within a framework of temporal and spatial absolutes, we now had a cosmos of relativity. Atoms that were supposedly indivisible were now divisible and changeable, no longer the unalterable building blocks of nature. The full implications of these findings and the total radical shift in thinking did not emerge until after World War I with the rise of quantum mechanics; however, the rumblings of these changes began in the Victorian period.

Scientists who were interested in psychical research or who had a commitment to spiritualism realized that these new findings in physics demonstrated how much we still did not understand natural phenomena. In particular, the exact relationship between mind and matter was unknown. Thus, a small but significant minority of physicists was drawn to investigating the various phenomena of spiritualism in the hopes that this might yield insight into the nature of the mind. Physicists began addressing themselves to questions that had previously been the domain of philosophers and theologians, including the possibility of human immortality.

At the same time, some of the most distinguished members of the physics community including John Tyndall, Michael Faraday, and Lord Kelvin rallied and brought all their logic and derision to discredit the spiritualist claims. These men had been responsible for revealing some of the great physical laws that they insisted were universal and uniform. They argued that repeatability of experiments was crucial and therefore spiritualism violated the basic tenets of good scientific practice.

Tyndall articulated a common criticism of the "pseudosciences." Science appeals to uniform experience, but the spiritualist phenomena were erratic. Committed spiritualists not only claimed that conditions were not always conducive to observe the phenomena, but attacked the very idea of uniform experience. They claimed there was no way to know whether uniform experience would continue to be so. However, scientists also recognized this. One of the most ironclad laws was Newton's law of universal gravitation. Nevertheless, as Thomas Huxley pointed out, the claim that a stone always falls to the ground because of the law of gravitation was ultimately based on experience. In all of human experience, stones have always fallen to the ground, and there was no reason to believe that tomorrow a stone would not fall to the ground. Stones fall to the ground was a "law of nature," but Huxley balked at making a statement that the stone *must* fall to the ground. Nevertheless, Huxley also recognized that the whole practice of science was premised on the belief that we can discover uniform laws of nature. Such a belief has been an extremely powerful and effective principle in generating new knowledge resulting in all kinds of practical applications. The claims of spiritualism could not boast such achievements.

In addition, much of the spiritualist phenomena could be explained in other ways. Michael Faraday claimed that the table turning that often occurred at séances was due to the unconscious effort of the sitters, "the mere mechanical pressure exerted inadvertently by the turner." Faraday had no use for mesmerists or phrenologists who used terms such as electrobiology or electrophysiology, cloaking their claims in scientific language. He argued that while they attributed unusual phenomena to electricity and magnetism, they had no understanding of the laws governing these forces, which made their claims ridiculous. They also claimed that these phenomena were due to some new undiscovered physical force, but they had not seriously investigated whether known forces were sufficient to explain the phenomena. Kelvin was especially dismissive, lumping together animal magnetism (mesmerism), table turning, spiritualism, and clairvoyance, describing it all as "wretched superstition." Tyndall admitted that he sometimes crossed the boundary of experimental evidence in his scientific speculations, but believing in the continuity of nature, he did not think it was necessary to bring in the occult to explain phenomena. He boldly asserted that science would eventually wrest from theology the entire domain of cosmological theory. By making such a claim, he was suggesting that a scientist had a right and even a duty to supplement empirical findings with conjectures from his imagination.

However, many scientists were not convinced that a strict materialist view of the world adequately explained everything. As Professor Balfour Stewart argued, there were "strong grounds for supposing that our environment is something very different from that in which the atomic materialists would wish to confine us."[23] The Society for Psychical Research was founded in 1882 and was in a different class than the other spiritualist organizations, having a membership of many prominent scientists including Nobel Prize winners John Rayleigh, J. J. Thomson, and William Ramsay. At its peak, it had over nine hundred members, including honorary members William Gladstone, John Ruskin, and Alfred, Lord Tennyson.

The society investigated a variety of controversial mental phenomena and produced reports of high caliber. Society members attempted to make links with more orthodox branches of psychological research. Just like Tyndall, they applied their imaginative and critical thinking to the data of physics and chemistry, but came to very different conclusions. They were united in their belief that psychic phenomena deserved serious scientific inquiry in that it might provide new insights into the working of the universe. However, it would not be quite true to claim that these men were committed spiritualists. Most of them remained skeptical; however, they also admitted that there were phenomena at séances that they considered genuine, but could not adequately explain.

Most of these scientists did not believe in the existence of spirits who were responsible for sending the messages. Rather, they were interested in the possibility of discovering some new psychic force and developing a theory of the cosmos that both science and theology could accept. In *The Unseen Universe or Physical Speculations on a Future State*, Stewart and P. G. Tait postulated the existence of an invisible world that science did not have the means to investigate, but this unseen world was linked to the visible world by bonds of energy. Modern physics was moving in just such a direction and later would confirm such ideas.

Stewart also suggested that intelligent beings besides humans possibly existed, and he hinted at the workings of "electrobiology." Using such language was exactly what Faraday had criticized. However, Stewart was merely making use of scientific jargon to describe the mesmeric trance. He was not using the term to impress people, but rather out of genuine perplexity. He loathed materialism, and he wanted to justify his faith in God and immortality, but as a scientist he found the most effective language to express his ideas was that of scientific naturalism. The concepts of energy and electricity had a physical basis that the spiritualist hypothesis lacked. Since so much was not understood about the nature of electricity and energy, it was not unreasonable to think that such forces could be useful in elucidating psychical occurrences. Like Crookes, he was searching for a satisfactory naturalistic explanation for what was considered supernatural phenomena.

In spite of the enormous amount of chicanery surrounding spiritualism, it was not obvious that spiritualist phenomena should be excluded from what were considered legitimate topics of scientific investigation. The scientific profession could not object to the idea of unknown or known forces working in mysterious ways. Electromagnetism when it was first discovered was just such a force. An undiscovered force could potentially explain psychic phenomena without recourse to spirit agents. As phrenology became discredited scientifically and became essentially a social movement for self-improvement, Victorians still were looking for other scientific explanations for the nature of mind. Darwin's theory of evolution offered another approach to the study of mind. Indeed, Darwin ended *The Origin of Species* by claiming, "psychology will be based on a new foundation. . . . Light will be thrown on the origin of man and his history."[24]

Darwin's words were prophetic. Most of the important psychological research that was done in the last part of the nineteenth century was heavily informed by evolutionary assumptions. This continued to be even truer in the twentieth century. Psychologists interacted with biologists, zoologists, and anthropologists, many of whom worked within a framework of evolutionary thinking. Just like the developments in physics, Darwin's theory both undermined and gave hope to providing scientific evidence for the traditional Christian belief in an afterlife. Many spiritualists and psychic researchers used evolutionary theory to bolster their claims. Nevertheless, such claims were problematic. The complexities surrounding spiritualism, evolutionary theory and the origin of the human races are dramatically illustrated in the life and work of Alfred Russel Wallace. The next chapter explores the impact of evolutionary theory on spiritualism and psychic research.

5

Morals and Materialism

Alfred Russel Wallace, Spiritualism, and the Problem of Evolution

> The line between biology, morals, and magic is still not generally known and admitted.
>
> —C. D. Darlington, *The Facts of Life*

IN THE PREVIOUS chapter I argued that spiritualists and psychic researchers were particularly interested in the relationship of mental to physical phenomena because an independent existence of mind from the tissues of the brain was essential to their argument against materialism. Furthermore, most Victorians thought that the basis of morality depended on a system of religious beliefs that included the certainty that the mind was more than just a collection of molecules and chemical reactions. Nevertheless, the success and growing prestige of science meant that natural theologians, spiritualists, and psychical researchers alike were all drawn to a scientific perspective in their study of the mind. Ironically, one of the legacies of the scientific revolution was a Cartesian dualism separating matter from mind that effectively prevented the empirical investigation of the relationship of the mind to the brain. Questions of mind were left to the philosopher/psychologists to be solved by introspection and speculation. However, as chapter 3 pointed out, the monopoly that philosophers had on the study of mind and the nature of consciousness was challenged. Gall argued that the brain was simply the organ of the mind. He along with researchers in the developing disciplines of physiology, neurology, and psychiatry promoted the study of the mind from a biological perspective. In particular, Darwin's theory of evolution suggested a research program for the study of mind with far reaching implications. Thus, it is worth looking at the development of Darwin's theory in some detail. Darwin and Alfred Russel Wallace shared certain similar experiences that led to them independently to come up with the theory of natural selection. However, there were important differences in their background that shaped the interpretation of those experiences and eventually led to Wallace breaking with Darwin over natural selection as a complete explanation for the evolution of humans. Once again, "facts" are not disembodied entities in nature waiting to be discovered, but rather are observations that are interpreted within a cultural, social and political matrix at a particular historical moment.

Charles Darwin and the Development of His Theory

Charles Darwin was born into a wealthy English family on February 12, 1809. His father was a physician, and his mother came from the Wedgwood family, which had become quite well known as well as rich for its pottery and china. His grandfather Erasmus had propounded a theory of species change similar to Lamarck's. At an early age Charles showed an interest in nature and became an avid collector of all sorts of things—shells, seals, coins, and minerals. He was not a particularly brilliant student, causing his father to chide him, "You care for nothing but shooting, dogs, and rat-catching, and you will be a disgrace to yourself and all your family."[1] Following the family tradition he was sent to University of Edinburgh to become a physician. He, however, found the lectures "intolerably dull," was nauseated by the operating theater, and soon dropped out. His father decided he should become ordained in the church, and he entered Christ's College in Cambridge in 1827. This did not seem like a bad alternative to the young Charles. As a country parson he would have enough leisure time to pursue his interest in natural history and indulge in his favorite sport of shooting game. As with Edinburgh, he disliked most of his official studies remarking "no one can more truly despise the old stereotyped stupid classical education than I do." Instead, he spent most of his time studying natural history. He had become friends with the zoologist Robert Grant, a Lamarckian, in Edinburgh and in Cambridge he soon made the acquaintance of John Henslow, a professor of botany, and Adam Sedgwick, a professor of geology. While at Cambridge he also became a passionate collector of beetles. In addition, he was deeply influenced by William Paley's *Natural Theology.* He had read Alexander von Humboldt's accounts of travels and became excited about the possibility of studying natural history in the tropics.

He learned that the navy was sending a small ship, the HMS *Beagle*, to chart the waters of South America. The captain, Robert FitzRoy, wanting a gentleman companion to relieve the monotony of the voyage, had created the position of ship's naturalist to describe the areas visited, and Darwin was offered the job. Charles was elated, but his father was not. He thought that the voyage would be disreputable to Charles's reputation as a clergyman, that it was a wild scheme, and that Charles would never settle down. Fortunately for Charles, his Uncle Josiah Wedgwood intervened. On December 27, 1831, Darwin set sail on a five-year voyage that would change his life and start him thinking about ideas that would profoundly affect Western civilization.

Henslow had recommended that Darwin take with him the first volume of Charles Lyell's *Principles of Geology*; the second reached him while in South America. But Henslow had also cautioned Darwin, "on no account to accept the doctrine there espoused"—a warning not heeded. At the time, the prevailing view of earth history was that of catastrophism. The earth had been shaped by a series of catastrophic episodes, events that had no counterpart in the present day. Lyell revived James Hutton's uniformitarianism and argued that the same forces that shaped geological formations in the past were in operation today. The present was the key to understanding the past, and the young Darwin had ample opportunity to put this precept into practice. While in South America he was able to make several extensive excursions inland and in 1835 he observed the terrible effects of a major earthquake in Chile. This devastating upheaval had permanently raised the level of some land around Concepčion by as much as ten feet, and Darwin used this fact to explain how a vast number of fossil sea shells at Valparasio could have ended up at elevations of over twelve thousand feet in the Andes. Over long periods of time such movements would produce large-scale effects such as the building of mountain ranges.

Principles of Geology provided a totally naturalistic account of earth history. No supernatural interventions such as the biblical flood were necessary. The *Principles*, however, was not just a book of geology. Lyell realized that the series of geological strata and the sequence of forms in the fossil record were different aspects of the same problem. Were species real and permanent, or were they capable of being indefinitely modified over a long period of time? In a detailed critique of Lamarck's theory of transmutation, he concluded that the evidence did not support Lamarck's views. Nevertheless, *Principles* in reality gave a mixed message about the transformation of species. Lyell admitted it was often difficult to see the boundaries between species and varieties, with some varieties differing more among themselves than what were considered distinct species. His discussion of the distribution, the extinction, and the creation of species, and of competition, and variation was just the type of evidence that Darwin would use to support his own theory of evolution by natural selection.[2] Darwin adopted Lyell's idea of slow gradual change in the inorganic world and applied it to the organic world.

As the *Beagle* continued its voyage, it stopped for some time at the Galapagos, a group of relatively young volcanic islands several hundred miles off the coast of Ecuador. Darwin collected extensively from several of the islands and it soon became apparent to him that the islands had been colonized from the mainland of South America. Yet, the present-day inhabitants were distinct species from those of the mainland and often distinct from each other as well. As Darwin was leaving, he was told that the local people could tell which island a particular giant tortoise came from just by looking at its shell. Other organisms, such as mockingbirds and finches also appeared to be similar, but distinct from the mainland species and from each other. The full implications of these findings did not become apparent to him until later, but it certainly was suggestive that as the islands became colonized, the island inhabitants diverged from their ancestors on the mainland, eventually becoming distinct species. Darwin's observations on the Galapagos along with other findings such as the geographical distribution of organisms elsewhere and the geological formations led him to ponder what he had seen. Darwin began to think species were not immutable, but rather that they had gradually become modified.

However, a key question remained: how did such change occur? Although it was apparent to Darwin that species had changed over time, it was also apparent that the Lamarckian mechanisms of change—the action of the environment and the will of the organisms—did not provide an adequate explanation of how organisms were so beautifully adapted to their habits of life. He had been profoundly impressed with William Paley's ideas, and he saw firsthand the evidence of nature's beautiful adaptations: from the differently shaped beaks of the finches to the hooked and plumed seeds to aid in seed dispersal. He had always been "much struck by such adaptations, and until these could be explained it seemed to me almost useless to endeavour to prove by indirect evidence that species have been modified."[3] Nevertheless, the subject haunted him.

In 1837, Darwin started to systematically assemble ideas and evidence in a series of private notebooks. The project soon expanded to include not only the origins of plants and animals, but humans as well. He wrote to people living abroad such as missionaries to give him their impressions of native people. He went back and examined his specimens from the Galapagos more carefully—no small trick, as the finches had not been labeled from which island they came from. Quite early on he recognized that the familiar practice of plant and animal breeding might offer clues to the origin of species. He knew that ultimately domesticated plants and animals had come from wild ancestors in a process involving descent with modification. The breeders continually selected individuals with desirable traits and then bred those individuals among themselves. All the different varieties of dogs, from an Irish wolfhound to a cocker spaniel, had descended from common stock, the result of selective breeding. Indulging in a hobby common to many English gentlemen, Darwin became a bit of an expert pigeon breeder and realized that given enough time and patience artificial selection could cause astonishing results. Could a similar process be occurring in the wild?

In October 1838, Darwin happened to "read for amusement" Thomas Malthus's *Essay on the Principle of Population.*[4] This essay provided the missing piece to the puzzle of species change. Malthus argued that there was a universal tendency for human populations to greatly exceed their available food supplies, with the result that there was a continual "struggle for existence." Urban overcrowding, intense competition for jobs, poverty, disease, all supported Malthus's principle of population. One only had to visit the east end of London to see that life for its poor inhabitants was quite literally a struggle for existence. Darwin immediately realized that this struggle was not confined to human populations. He knew from his own experiments that far more individuals were born than survived to reproduce. Who survived and who perished? Was it just a matter of chance? To the casual viewer the individual seedlings in a forest or the offspring in a population of robins might look identical, but Darwin had observed that the amount of variation within populations was essentially limitless. Darwin reasoned that useful variations would tend to be preserved and unfavorable ones would be destroyed in the ongoing continual struggle for survival. Those favorable variations would accumulate through the generations in the same way as desirable traits accumulated in domestic populations that were the product of selective breeding. Eventually a new species would be formed. With the analogy of artificial selection in mind Darwin called this process natural selection. Natural selection not only provided a mechanism for species change, but it explained how species change *adaptively* in the continual struggle for existence.

Such a mechanism, resonating with ideas that were commonplace in nineteenth-century England meant that Darwin's theory would not be of interest to just men of science, but to all of society. Each group with its particular agenda found that Darwin's theory was infinitely pliable and could be used as evidence for its own point of view. Capitalists and socialists both claimed justification for their ideas in the tenets of Darwinism. Spiritualists also bent Darwinism to their own needs.

Spiritualism and Evolutionary Theory

Much of the general public, embracing the idea of struggle and adaptation, applied it to humans and believed that the person who had prevailed in this struggle for survival was the better person. Psychical researchers were also drawn to the idea of the "better" person, but they did not uncritically accept the unbridled optimism that characterized the segments of society that had been benefiting from the rapid industrialization. The record of human progress was ambiguous, particularly in relation to morals and ethics. Still, many people hoped that the evolutionary process offered the possibility of a promising future for humankind, and they looked to the study of the human mind for supporting evidence. The majority of British spiritualists were more than willing to embrace the idea of organic change because Darwin's idea confirmed their own belief in progressive development beyond the veil. While Darwin emphasized where we had come from, the spiritualists were more interested in where we were going and believed that we were evolving to "progressive spiritual states in new spheres of existence" after physical death. In addition, evolution in the spirit world proceeded at a more rapid rate in conditions that were more favorable to growth.

Many religious people were attracted to spiritualism because it kept the belief in an afterlife intact while allowing them to jettison the repressive beliefs concerning humankind's fall from grace. Evolution was seen in aiding their cause. For them, evolution meant that all things progressed upward into higher forms of life. Man did not begin perfect and end in a "fall." He began imperfect, and was steadily going on in an onward and upward path, out of the animal's darkness into the angel's marvelous light. In the afterlife evolution continued, but only in the sphere of the mind.

Many spiritualists saw no incompatibility between Darwin's theory of evolution and spiritualist theories concerning human development before and after death, but their knowledge of biology was quite limited. They were not in a position nor did they have any desire to evaluate critically the popular accounts of evolutionary theory. However, for someone like Alfred Russel Wallace who had independently of Darwin developed the theory of natural selection, and understood only too well the implications for human development, to combine evolution with spiritualism was far more difficult. If natural selection was solely responsible for "mans place in nature," then humans, like all other organisms had evolved from some earlier organism. If humans had evolved from some other prehuman organism, they were not qualitatively different from other organisms. The difference was only a matter of degree, not kind. The human brain had merely undergone further development than that of other organisms. If the human brain was merely more developed, then there was nothing basically distinctive about the human brain. There was nothing unique about humans, neither mind, nor consciousness, nor spirit. Rather than supporting a vision of progress in the hereafter, it seemed that evolution undermined the most basic tenets of spiritualism. Yet, Wallace managed to reconcile his belief in evolution by natural selection with an absolute faith in spiritualism, and in doing so, his life exemplifies many of the hopes, fears, and contradictions of Victorian society.

Alfred Russel Wallace

Alfred Russel Wallace was born in the Welsh border village of Usk in 1823. He spent a happy if somewhat impoverished childhood. He described his formal education as worthless, except for learning enough Latin to later understand the names and descriptions of species. Far more interesting to him were his own readings at the town library where his father had become the librarian. His older brother John and he spent many happy hours in a loft of a small stable building all kinds of toys, fireworks, mechanical devices, and even a small cannon, making use of whatever odds and ends that could be found or bought for a few pence. This resourcefulness would serve him well in his later travels in South America and the Malay Archipelago.

In 1837, Alfred was sent to London to live with John, who was apprenticed to a master builder. Although he was just thirteen and spent only a few months there, this time profoundly influenced him. For it was here that his brother took him to evening meetings at the Hall of Science, and he imbibed the social and political philosophy espoused by Robert Owen. This was the beginning of the radical and egalitarian ideas that characterized his mature political and social thought. He attended lectures on secularism and read Thomas Paine's *The Age of Reason*. Christianity did not seem to deal adequately with the problem of evil. "Is God able to prevent evil, but not willing? Then he is not benevolent. Is he willing, but not able? Then he is not omnipotent. Is he both able and willing? Whence came evil?"[5] Other readings provided a totally convincing argument against the idea of eternal punishment. The young Alfred abandoned the traditional Christian teaching of his childhood and became a thoroughgoing freethinker, concluding "that the orthodox religion of the day was degrading and hideous, and that the only true and wholly beneficial religion was that which inculcated the service of humanity, and whose only dogma was the brotherhood of man."[6]

In the summer of 1837, Wallace joined his brother William in Bedfordshire, to learn surveying and mapping. The work allowed him ample time to wander the hills and it was in these ramblings that he discovered his passion for nature. "I experienced the joy which every discovery of a new form of life gives to the lover of nature, almost equal to those raptures which I afterwards felt at every capture of new butterflies on the Amazon."[7] He bought some books on botany and began collecting the flora of the area. It was while working as a surveyor that he saw the hardship and poverty caused by the Land Enclosure Act, making it virtually impossible for people to eke out a living. Wallace would later become a fervent advocate for the abolishment of private property, calling for the nationalization of all lands.

In 1844, he read two works that were to be crucial to his career just as they had been to Darwin's: Alexander von Humboldt's *Personal Narrative of Travels* and Malthus's *Essay on the Principle of Population.* He had earlier read Darwin's *Voyage of the Beagle,* and this combined with Humboldt's vivid descriptions of the tropics of South America made him determined to find a way to visit the area. The significance of Malthus would come much later, just as it had with Darwin, after they had both traveled and made extensive observations around the globe. In Leicester, Wallace met Henry Walter Bates, an entomologist and they soon found that they were kindred spirits. Like Darwin, Bates also had a passion for beetles and soon enticed Wallace into becoming a collector of insects as well as plants. However, Wallace's curiosity was not just confined to shrubs and bugs. In these other pursuits are clues to his adoption of spiritualism in his later years.

Wallace became quite interested in mesmerism, attending lectures and demonstrations, and even succeeded in putting some boys into the trance-like state. He soon became convinced of mesmerism's legitimacy. He read Combe's *The Constitution of Man*, which resonated with his own political views and tapped into his growing interest in psychical phenomena. He became a strong advocate of phreno-mesmerism. "I established to my own satisfaction the fact that a real effect was produced on the actions and speech of a mesmeric patient by the operator touching the various parts of the head; that the effect corresponded with the individual expression of the emotion due to the phrenological organ situated in that part."[8] Like many other people in his generation, his attraction to these doctrines must be viewed within the context of his abandonment of orthodox religious views. His experiences of the next decades moved him further and further away from any traditional belief in God. Years later, he returned to a belief in a higher power, but at this time all of his readings propelled him to a position of agnosticism.

Wallace had also read William Swainson's *Treatise on the Geography and Classification of Animals*, which tried to reconcile scripture, geology, and zoology. But Wallace was unimpressed, writing in his own copy, "a most absurd and unphilosophical hypothesis," adding "to what ridiculous theories will men of science be led by attempting to reconcile science to scripture."[9] He had already decided that the doctrine of special creation could not possibly be true and believed that the problems of natural history would be solved by natural laws, not natural theology. The arguments in the *Vestiges* would find a very receptive reader in Wallace. He believed that the idea of natural law as presented in the *Vestiges* provided the correct path to understanding the natural world. Darwin, in spite of his disparaging remarks in letters to colleagues concerning the *Vestiges,* recognized much to his chagrin, that the anonymous author had used much of the same kind of evidence that he had used in developing his own theory of evolutionary change.

In 1848, Wallace and Bates realized their dreams to visit the tropics of South America. Once there they traveled together and separately, exploring and collecting. In 1850, at the convergence of the Amazon and the Rio Negro, he and Bates separated permanently, Bates exploring the Upper Amazon for the next eight years while Wallace explored the Rio Negro, Uaupés and other northern tributaries. Returning to London two years later, Wallace made the acquaintance of the leading lights of British science including Lyell, Darwin, and the up-and-coming Thomas Huxley.

After two years, Wallace left again in 1854 to the Malay Archipelago, where he lived for eight years and gathered over 125,000 specimens. In organizing them he pioneered the study of the geographical distribution of plants and animals, which later provided crucial evidence to advance his theory of evolution by natural selection. Wallace noticed a division within the flora and fauna of the Australasian islands, which followed a line of demarcation that today is still known as the Wallace Line. His massive two-volume *The Geographical Distribution of Animals* (1876) entitles him to be considered the founder of the science of zoogeography, and his book *The Malay Archipelago* (1869) remains one of the finest scientific travel books ever written.

Wallace was also the first prominent anthropologist to live for an extended period among the native peoples whom he studied. He seemed remarkably free of the racism and stereotyping of indigenous people that characterized the thinking of Darwin and virtually everyone else who lived at the time. Rather, he displayed a sympathy and understanding of these cultures that was far ahead of his time. He was impressed with their honesty, intelligence, aesthetic sense, and harmonious way of life. He was struck by the absence of a system of private property, which brought back to him the socialist teachings of Owen from his youth. When he returned to England he became increasingly critical of the excesses wrought by competitive industrial capitalism. In many ways the "savages" appeared to be more civilized than the so-called civilized races.

> Each man scrupulously respects the rights of his fellow, and any infraction of those rights rarely or never takes place. . . . There are none of those wide distinctions, of education and ignorance,

> wealth and poverty, master and servant; . . . there is not that severe competition and struggle for existence or for wealth, which the dense population of civilized countries inevitably creates. All incitements to great crimes are thus wanting, and petty ones are repressed partly by the influence of public opinion, but chiefly by that natural sense of justice and of his neighbor's right which seems to be, in some degree, inherent in every race of man.[10]

Wallace's view of native people provides an important clue to his later conversion to the spiritualist hypothesis.

Although Wallace was living in the remote jungles of Malaysia, his researches, which he regularly wrote up, had garnered him a reputation as a first-rate naturalist. But what brought him truly into the spotlight was his independent discovery of the theory of natural selection. In 1855 he wrote a brilliant essay, "On the Law Which Has Introduced the Introduction of New Species" that drew on his observations not only in the Malay Archipelago, but from South America as well. He had noticed that many species on the opposite sides of the rivers in South America were closely related, but not the same. Yet physical conditions were virtually identical. The geographical distribution of organisms on the Galapagos Islands had made a profound impression on him, just as they had on Darwin. On the theory of special creation one would have expected the organisms on the islands to be identical to one another since their environments were the same. They should have been significantly different from the mainland species since the environments were so different. Yet, each island had species unique to itself and from the mainland species, but nevertheless, they were similar in form. How did this happen? Wallace reasoned, as had Darwin that the islands must have been initially populated by species from the mainland through winds and ocean currents. Sufficient time had passed that the original mainland species had died out and only the modified prototypes remained. Rather than special creation, Wallace asserted "every species had come into existence coincident both in space and time with a preexisting closely allied species." This explained a wide array of observations including the geographical distribution and the often seemingly peculiar anatomical structures of organisms around the globe.

Wallace continued to gather data, all of which reinforced what he had observed on the Galapagos and in South America. In the Malay Archipelago, the organisms found in New Guinea and on the Aru Islands were completely different from the Western Islands such as Borneo. However, the two countries were virtually identical in terms of climate and physical conditions. In contrast, physical conditions could not be more different between Australia and New Guinea, but the two faunas were strikingly similar. Wallace concluded that something other than the law of Special Creation must be responsible for both the structure and distribution of organisms. Like Darwin, Wallace also searched for a mechanism of evolutionary change. While ill with malaria on the island of Ternate, he was rereading Malthus, and just as with Darwin, Malthus supplied the key insight for Wallace to independently come up with the law of natural selection.

Wallace had been corresponding with Darwin for years, but in 1858 he sent Darwin a letter along with his paper entitled "On the Tendency of Varieties to Depart Indefinitely From the Original Type" in which he outlined virtually the identical theory of natural selection. Darwin was devastated and wrote to Lyell, "I never saw a more striking coincidence; if Wallace had my MS sketch written out in 1842, he could not have made a better short abstract!"[11] Lyell and Hooker did not want Darwin to lose priority since he had outlined these very ideas many years earlier. They arranged for extracts from Darwin's unpublished writings and Wallace's paper to be read before the Linnaean Society. Wallace did not seem to mind this joint presentation and assured Darwin that he considered the theory of natural selection to be Darwin's theory only. He continued to defer to Darwin both publicly and privately, but his actions were somewhat curious. In later years he acknowledged on several occasions that he had independently discovered natural selection. Thus, it was partly due to Wallace's own actions that evolution by natural selection became known as Darwinism rather than the Darwin/Wallace theory of natural selection. There were significant differences between the two men's ideas, which became greater over the years. But in 1858 the similarities far outweighed the differences. Wallace stayed on for several more years in the Malay Archipelago, while Darwin frantically worked on a shorter version of his ideas and in 1859 published *The Origin of Species*. *The Origin* would create a storm of controversy, and that controversy ultimately revolved around the implications of evolutionary theory for humans.

Wallace, Human Evolution, and the Limits of Natural Selection

While corresponding with Darwin, Wallace had raised the question of human origins. Darwin replied "I think I shall avoid the subject, as so surrounded with prejudices, though I full admit that it is the highest and most interesting problem for the naturalist."[12] However, Wallace came to believe in species transformation by means of natural selection, not just by studying plants and animals, but also through his interest in ethnology and the origins of humans.[13] When the joint papers of Wallace and Darwin were presented before the Linnaean Society in 1858, we must assume that, like Darwin, Wallace did not think that human evolution was an exception or different from the process as it occurred in the rest of nature. Indeed, his thoughts on human origins were a great help to Darwin in his own thinking on the matter.

The publication of *The Origin* had raised the problem of human origins to new heights of controversy. In particular, it rekindled the eighteenth-century controversy over the origin of the different human races. Were human races separate species or just varieties of a single species? With the possibility that humans might actually be related to other primates, the monogenist-polygenist debate became reformulated. Were present-day human races derived from distinct species of apelike ancestors or just varieties that developed later from one common ancestor? In 1864 Wallace published a provocative paper entitled "The Origin of Human Races and the Antiquity of Man Deduced from the Theory of Natural Selection," which offered a subtle and wonderful compromise between the polygenist and monogenist hypotheses. In addition, it explained why the human body with the exception of the skull looked so similar to the bodies of the present-day apes, while at the same time the skull and mental abilities diverged widely from these same apes.[14]

Wallace suggested that humans, unlike other organisms, had experienced two distinct stages in evolution. In the first stage, human ancestors had evolved just as any other creature according to the principle of natural selection. Thus, individuals who carried favorable variations in the constant struggle for existence survived, their form continuing to be modified. But once the human brain had evolved to a certain point and the moral and intellectual capabilities were fairly well developed, natural selection would no longer act on the human physical form and structure, and it would remain essentially unchanged.

Humans were able to adapt to harsh and changing environmental conditions by the use of their intellect alone. A change in climate might result in an animal developing a thick coat of fur or a layer of fat, but humans would put on warmer clothes and build a shelter. No change in body structure would be required. A change in the abundance of food might result in an animal changing its diet or bodily weapons such as claws and teeth, or the digestive system might be altered. Humans, however, in an early period had already learned to hunt. Improved trapping and hunting techniques would allow them to survive with no change in body structure. As soon as humans possessed fire many new foods became edible, increasing the available food supply. Once they had developed agriculture and the domestication of various animals, the amount of food was also greatly increased. By intellect alone with no change in the body, humans could survive and adapt, remaining in harmony with the ever-changing environmental conditions. Furthermore, the developed social and sympathetic feelings allowed humans to band together for mutual comfort and protection, and such traits would be preserved and accumulated. Thus, Wallace believed that the more intellectual and moral races would replace the lower ones. Such views also suggested that early humans must have first arisen when the brain size was still fairly small, at a far earlier time than most people thought, and when natural selection would still be acting on the body.

Wallace's two-stage theory of human evolution provided a solution to the monogenist/polygenist debates. The polygenists claimed that evidence from ancient Egyptian tombs indicated as much difference existed between the Negro and Semitic races then as now. Such evidence contradicted the monogenist hypothesis that the further back in time one went, the more similar the races became. They claimed that if it was possible to go back to the time when humans made their first appearance as a species, only one race existed. Wallace agreed with the monogenists in this regard and did not find the Egyptian evidence very compelling for two reasons. First, he believed that humans had been around far longer than five thousand years. He suggested that humans might have existed as far back as the Miocene. Although at the time evidence did not exist for such a position, the incompleteness of the geological record provided a reasonable explanation for the absence of such evidence. Second, at a far earlier time, racial divergence had already ended because natural selection was no longer acting on the physical structure of these ancestral forms. Once the brain had reached a certain level of development, it protected the body from the action of natural selection. Only the mind continued to evolve. This would explain why little structural change would be observed between the races five thousand years ago and now, and would also explain why human bodies were so similar to ape bodies.

This last point also addressed the real question that underlay the monogenist/polygenist debate, which was not a scientific one about human origins, but a more insidious one concerning the intellectual and moral qualities of the different races. If the polygenists were correct that each race had a separate origin, then they were in reality different species. From there it was a small step to claim that some races were superior to others, which in turn was used to justify slavery, imperialism, and various racial theories. Monogenists, in arguing for a single origin of all human beings, claimed that they had scientific proof that the "savage" races were not qualitatively different or inferior to their masters or colonial administrators. Abolitionists and members of other humanitarian reform movements used the monogenist position to further their own goals.

Wallace's ingenious solution to this controversy essentially amounted to a moral monogenesis and a physical polygenesis. He argued that at a time far back in history humans indeed were a single homogenous race. At that time, our ancestors had a human form, but did not yet have a well-developed brain. They did not have speech or the sympathetic and moral feelings that today characterize the human species. If one did not consider these ancestors to be fully human until the higher faculties were fairly well developed, then one could assert that there were many distinct races of humans. However, if one thought that our ancestors whose form was essentially modern, but whose mental faculties were scarcely raised above the brute, still deserved to be considered fully human, then all of humankind shared a common origin. Thus, Wallace did not deny that physical and racial differences existed, which could have been due to separate origins, but that all depended on how one defined the origin of the human species. The different races all had evolved to a certain point in which they all exhibited a shared common humanity. If one defined the origin of the human species at that point in time, then the different races represented variants of a single species.

Darwin was very impressed with Wallace's paper, as he too had been contemplating the problem of human evolution. Claiming that Wallace was already far ahead of him in his thinking about the matter, Darwin offered to give Wallace his own notes on man, which were "in a state of chaos." Wallace declined the offer. That is where matters stood for about five years with Wallace and Darwin regularly writing each other, exchanging and developing their views on their joint theory. However, in 1869 in a review of new editions of Lyell's *Principles of Geology* and *Elements of Geology* appearing in the *Quarterly Review of London*, Wallace did a dramatic turnabout. Wallace now claimed that neither natural selection nor a general theory of evolution could "give an account whatever of the origin of sensation or conscious life . . . the moral and higher intellectual nature of man is as unique a phenomenon as was conscious life."[15] Darwin must have had some inkling that something might be amiss as he wrote to Wallace, "I shall be intensely curious to read the *Quarterly*. I hope you have not murdered too completely your own and my child."[16] Unfortunately for Darwin, his fears were soon realized.

In *The Origin* Darwin had emphasized and Wallace agreed that natural selection was a principle of utility. Natural selection could not preserve a harmful structure and furthermore, it was a principle of present utility and relative perfection only. A structure would not be preserved by natural selection because it would become valuable for future generations nor would natural selection accumulate favorable variations to provide a more perfect structure if a less efficient one could do the job in the ongoing struggle for survival.[17] Although other factors played a significant role in Wallace's change of view, he grounded his argument for the insufficiency of natural selection on the principle of utility alone.

Wallace reviewed what was known about the prehistoric races and also the present-day "savage" races and concluded that they had several traits that could not be explained by a utilitarian analysis. In particular, several aspects of humankind's intellectual and moral nature seemed inexplicable in terms of natural selection. Why would prehistoric man have needed such a large brain? In addition, three other physical features could not be explained by natural selection: the hand, the loss of hair, and the organs of speech. All of these characteristics could not be accounted for because they were present in the prehistoric and "savage" races in a much more highly developed form than was necessary for the struggle for existence. The loss of hair was even harmful; therefore, how could it have come about by natural selection? However, these traits were all useful for civilized man and this suggested to Wallace that they were a provision for future use. If they were a trait intended for the future, they could not have come about by natural selection.

Wallace did not accept the racist thinking of virtually everyone at that time, recognizing that the "savage" races were no different from the civilized Victorian gentlemen. They were just as intelligent, had the same moral capabilities, and the same range of emotions. However, Wallace did not think it was necessary for them to be in possession of such faculties. They bore no relation to the wants, desires, or welfare of those people in the conditions they were living in. This would have been equally true for the prehistoric races. For Wallace, in the condition that these native people lived, their actual needs and wants would produce a brain "a little superior to that of an ape." Wallace now believed the "savage" had these capabilities in spite of their being of no use to him because this had been a provision for humankind's future. He applied the same argument to the structure of the hand. He claimed that the higher apes also had hands far more developed than they needed for survival. However, it would be necessary for humans to have such a perfect hand for later developments in the arts and sciences. There was no doubt that the "savage" races had the same vocal capabilities as the civilized races, but Wallace claimed they wouldn't need them. However, civilization absolutely depended on our power of speech. Again, this was a provision for future use, not present-day use. But natural selection did not anticipate the future. Wallace thought a variety of other features of external form were also useless, including erect posture, delicate expressive features, beauty and symmetry of form, and smooth naked skin. Nakedness was even harmful to "primitive" and prehistoric races and thus, once again, natural selection could not account for them. Wallace suggested that the delicate expressive features and beauty of form were necessary for the refined emotions and aesthetic sense that developed in civilized cultures. Naked skin would stimulate humankind's inventive faculties to devise clothing, which would develop their feelings of modesty and contribute to their moral nature. All of these traits seemed to be preparation for the future.

If natural selection could not account for these traits, then how did they come about? Wallace's answer surprised everyone. There was some power that was guiding the action of the "great laws of organic development" in definite directions for special ends. Ironically, Wallace used the same analogy of domestic variation for which he had criticized Darwin in *The Origin* to suggest that a higher intelligence was guiding the process of evolution. Just as humans selected particular variations to produce desirable varieties of fruits, vegetables, and livestock, an "overruling intelligence" had watched over the action of laws of variation and selection, directing the variations, which resulted in "the indefinite advancement of our mental and moral nature."[18] Darwin was horrified; marking his copy of Wallace's review with a triple underlined no followed by numerous exclamation points.[19]

Wallace insisted this was not a negation of the principle of natural selection. Rather, just as in artificial selection the process had been supplemented by the conscious selection of humans, the "laws of organic development have been occasionally used for a special end just as man uses them for his special ends." He continued to ground his argument on the grounds of utility alone, maintaining that human beings had a variety of unique traits that simply would not have been useful in the very lowest state of civilization. These included mathematical ability, ability to form abstract ideas, ability to perform complex trains of reasoning, aesthetic qualities, and moral qualities. All "savages" possessed these latent capabilities, but their needs did not require these abilities. Only rarely were they used. These people did not need to form or use abstract ideas because their language contained no words for them. They did not reason on any subject that did not appeal immediately to the senses or foresee beyond the simplest necessities. Wallace did not think that their ability to design and use weapons was any more ingenious than the jaguar he was hunting. Other animals seemed to have as much forethought as the "savage." Furthermore, he certainly did not need to have the ability to form ideal conceptions of space and time, of eternity and infinity, to have capacity for intense artistic feelings of pleasure in form, color, or composition. The native did not make use of geometry or other abstract ideas concerning form and number. "Natural selection could have endowed savage man with a brain a little superior to that of an ape, whereas he actually possesses one very little inferior to that of a philosopher."[20] Wallace even doubted whether civilized man made full use of these capabilities. He concluded that they were for the use of civilized man's future, not just for prehistoric man's future.

The most difficult trait to account for by natural selection was conscience, the development of a moral sense. This seemed to be totally inexplicable on the grounds of utility alone. Wallace claimed that the practice of honesty might be understandable, but the moral sense included more than that. Rather, it was the feeling of sanctity for such concepts as honesty. He did not see how the mystical devotion toward objects of moral worth could have been any aid to survival in the jungle. In fact, all through history untruthfulness has been allowable in love, even laudable in war. Most people didn't seem to value it at all when it came to trade or commerce. A certain amount of untruthfulness was necessary as politeness in both Eastern and Western culture, and even extreme moralists believed a lie was justifiable to elude an enemy or prevent a crime. Thus, it was hard to support the idea of honesty on a total utilitarian analysis. However, morality was an essential part of human nature. Wallace concluded that the feelings humans have about right and wrong came prior to any experience of utility, and since that was the case they could not be the result of natural selection. A superior intelligence guided the development of man in a definite direction and for a special purpose. For Wallace, the law of natural selection was insufficient to produce the "ultimate aim and outcome of all organized existence—intellectual, ever-advancing, spiritual man."[21]

Many people have claimed that Wallace "changed his mind" about human evolution and that change was due to his involvement with spiritualism. However, the story is not quite so simple. It is crucial to understand why Wallace was open to the claims of spiritualism while Darwin, Huxley, and various other members of the Darwinian camp were not. A variety of factors drew Wallace to the ideas of spiritualism. His working-class background, his exposure to the socialist ideas of Robert Owen, his disenchantment with traditional Christian doctrine, along with the years of wandering and solitude in the jungle, meant that he did not participate in the institutions of middle-class social and intellectual life. He had little empathy for the professional goals or the cultural values of most of his scientific colleagues. For the emerging class of professional scientists, natural selection represented a key feature of the naturalistic cosmology they were advocating. But Wallace regarded the investigations of nature quite differently. It was an aid for the moral development of human beings. Such a view suggests that Wallace may never have believed that natural selection provided an adequate explanation for the presence of the higher intellectual and moral qualities of humankind all along. His two-stage theory of human evolution left no doubt that it was the mind and moral faculties that turned the human animal into a human being. He was also quite optimistic about the future of the human race. His 1864 paper ended with a utopian vision for the future of humankind, which foreshadowed his later claim that eventually it would be the pure spirit of humans that would continue to evolve. As technology continued to advance and spread throughout the world, the mental and moral qualities would continue to develop and eventually a single homogenous race would eventually exist in which

> no individual . . . will be inferior to the noblest specimens of existing humanity. Each will then work out his own happiness in relation to that of his fellows; . . . the well balanced moral faculties will never permit any one to transgress on the equal freedom of others; restrictive laws will not be wanted, for each man will be guided by the best of laws; a thorough appreciation of the rights and a perfect sympathy with the feelings, of all about him . . . mankind will have at length discovered that it was only required of them to develop the capacities of their higher nature; in order to convert this earth, which had so long been the theater of their unbridled passions, and the scene of unimaginable misery, into as bright a paradise as ever haunted the dreams of seer or poet.[22]

In such a world, slavery and other kinds of repression would be abolished. Eventually, humans would evolve into a new and distinct order of being.

Wallace had a conception of human beings that drew heavily on the ideas of phrenology. Not only had phrenological doctrine permeated the *Vestiges*, but also as previously mentioned, much earlier he had read Combe's *The Constitution of Man*. He had attended lectures on the topic, purchased a phrenological bust, and had his own head examined on two different occasions. Both phrenology and mesmerism had a concept of mind that was compatible with spiritualism. The trance of a mesmerized subject was quite similar to the trance state of the medium. Phrenological doctrine taught that the brain was the organ of the mind, and many interpreted this to mean that mind was merely an epiphenomena of the brain. However, Wallace did not agree. He never claimed that brain and mind were identical. In addition, it was precisely those faculties that phrenology maintained were unique to humans such as veneration, conscientiousness, hope, and wonder that Wallace claimed became highly developed once the evolution of the brain shielded the human body from the action of natural selection.

One of the most heated scientific debates in the middle of the nineteenth century concerned the supposed progressive nature of the fossil record. Natural selection explained why species diverged over time, but did not explain why the change would necessarily result in organisms of increasing complexity. Thus, many people used the progressive nature of the fossil record as evidence for a revised version of the argument from design. But Wallace had rejected the creationist idea that organisms were adapted to their particular phylogenetic position, and he was also convinced that many human traits were *not* the result of adaptation to environmental conditions. Nevertheless, he was convinced that organisms were evolving toward higher and higher states of perfection. Thus, he did not regard his adoption of spiritualism as a change of mind or a regression, but rather as an extension to natural selection in giving a complete account of the history of life.[23] He insisted that his differences with Darwin did not take away from the overwhelming significance of natural selection.

Wallace continued to publish first-rate work in biogeography that made no reference to his spiritualist beliefs. However, once he had become convinced of the reality of spiritual communication and spirit manifestation, these phenomena demanded survival after bodily death and that within the human body an organized spiritual form existed. For Wallace, the essence of man was his spirit: "If you leave out the spiritual nature of man you are not studying man at all," and natural selection could not explain this spirit.

Wallace's Involvement with Spiritualism

Wallace appears to have shown no interest in spiritualism until the mid-1860s. In the preface to *Miracles and Modern Spiritualism*, he described himself at the age of fourteen as a "confirmed philosophical skeptic" who was so "thorough and confirmed a materialist" that he had no interest in spiritualist phenomena.[24] While wandering the tropics he had heard of the table turnings and spirit rapping occurring in the United States and Britain, but thought that many of the descriptions of the phenomena were so outlandish that they must have been the ravings of madmen. Others, while well authenticated and inexplicable according to the ordinary laws of nature, he, nevertheless, concluded that they "*must* be either imposture or delusion."[25] However, he also knew from his experience of mesmerism "that there were mysteries connected with the human mind which modern science ignored because it could not explain." Being a curious fellow, he decided to investigate the matter for himself as soon as he returned home. Reasonably quickly Wallace became convinced that the phenomena were genuine.

Wallace attended his first séance in July 1865 at the home of a skeptical friend with the family. No medium was present. Faint taps that gradually increased and became distinct were heard and the table moved quite considerably. He then experienced a curious vibration from the table, almost like the shivering of an animal. These phenomena occurred for two hours.[26] He continued to attend séances where he conducted a variety of "experiments" trying to elucidate the phenomena. He asked various people to leave until he was alone at the table and still there were taps and vibrations. Convinced that there was no deception he ended his notes: "These experiments have satisfied me that there is an unknown power developed from the bodies of a number of persons placed in connection by sitting round a table with all their hands upon it."[27]

In September, Wallace started to attend séances with Mary Marshal, a renowned British medium, usually accompanied by a good friend "of thoroughly skeptical mind." They witnessed a variety of physical and mental phenomena including table levitation and the movement of a guitar which "behaved as if alive itself." Communications occurred concerning the names, ages, and other particulars of relatives of those present at the séance. Wallace received a communication that spelled out the place where his brother had died, his Christian name, and at Wallace's request the name of Henry Bates who had last seen his brother alive. Wallace thought he had taken precautions against deception. He had examined the furniture beforehand in the séance room and had marked pieces of paper before spirit writing tests to preclude substitution when the lights went out.[28] Wallace was also reading the spiritualist literature as well. By the end of 1866 he was convinced that no physical force could elucidate what he had witnessed. Some unidentified impersonal psychic force was necessary to explain how the mediums could transmit information, to explain the passage of matter through matter, and to explain how objects acted in the defiance of gravity. He had a pamphlet printed at his own expense, "The Scientific Aspect of the Supernatural," which contained testimony of highly respected people describing the phenomena, leaving out his own experiences because he had not had any in his own home. But that would soon change.

Wallace started to attend sittings regularly with Miss Nichols, a friend of his sister and a medium who became quite famous as Mrs. Samuel Guppy. Again, exceedingly remarkable phenomena were observed. A small table rose. Another time one levitated to the chandelier even with the stout medium on it. To preclude the possibility that the medium had not just lifted the table with her foot, in a second trial without telling anyone Wallace had put some thin tissue paper between the feet, an inch or two from the bottom of the pillar so that any attempt to put some one's foot underneath would crush and tear the paper. The table rose and then dropped down, rose and dropped again. Wallace then turned up the table and "much to the surprise of all present, showed them the delicate tissue stretched across altogether uninjured!"[29] The most striking effect was the appearance of an apport. All of a sudden the tables were strewn with fresh flowers complete with drops of dew on petals and leaves. Their fragrance filled the room, so the flowers could not have been hidden in the room preceding the séance. He wrote in the *Spiritualist Magazine* "15 chrysanthemums, 6 variegated anemones, 4 tulips, 5 orange berried solanums, 6 ferns of 2 sorts, 1 *Auricula sinensis* with 9 flowers—37 stalks in all."[30]

Wallace had become a true believer and he tried to interest his scientific friends to attend séances. However, he was largely unsuccessful. He asked Huxley to come to the weekly sittings with Miss Nichols to observe the phenomena "before deciding that we are all mad."[31] Huxley replied that he "was not disposed to issue a Commission of Lunacy against you. It may even be all true." But Huxley simply wasn't interested. "I never cared for gossip in my life, and disembodied gossips such as these worthy ghosts supply their friends with, is not more interesting to me than any other."[32] Wallace replied that he was not interested in gossip either, but believed that he could demonstrate the existence of a force that has been declared impossible and of an intelligence from a source that had been deemed absurd. He pointed out that Michael Faraday had claimed that if such a force could be demonstrated that has not been recognized by science, this would certainly deserve applause and gratitude. Wallace was convinced that he could demonstrate such a force and that the scientific community should investigate it. However, Huxley never attended a séance with Wallace. Wallace attempted to interest various other colleagues including Darwin, the physiologist W. B. Carpenter and John Tyndall. Darwin had read some of Crookes's articles on his spiritualist investigations and felt he could not just dismiss spiritualism as he had mesmerism. His cousin Francis Galton had attended several séances hosted by Crookes with D. D. Home and Katie Fox. Galton had been quite skeptical, but had been impressed with some of the phenomena he had witnessed and urged Darwin to join him in an investigation of psychical phenomena with Home. That meeting never took place, but Darwin did attend a séance in 1874 with the medium Charles Williams. Darwin became hot and tired and went upstairs at the crucial moment. When he returned he saw that everything had been strewn all over and was told that the table had been lifted over the heads of those who had been sitting round it. While Darwin was perplexed, he nevertheless declared, "The Lord have mercy on us all, if we have to believe in such rubbish." A week later, Huxley, incognito, attended another séance with Williams. He wrote to Darwin, giving a detailed report of the séance describing the various phenomena, but also describing how they could have been produced. He concluded that Williams was "a cheat and an impostor." Darwin was relieved to receive such news and claimed that "now to my mind an enormous weight of evidence would be requisite to make me believe in anything beyond mere trickery."[33] Tyndall and Carpenter each attended one séance, heard some weak tapping, were unimpressed and never returned. Tyndall had already been quite critical of Wallace's pamphlet, writing to him that he had read the book "with deep disappointment." Although the logic of the book showed Wallace's sharp mind, he deplored Wallace's willingness "to accept data which is unworthy of your attention."[34]

The Darwinian Camp Responds

Darwin, Huxley, Carpenter, and Tyndall had no use for spiritualism, and they were amazed at Wallace's credulity concerning the phenomena he observed at séances. Wallace admitted that once he had become convinced of the reality of the existence of higher spirits, that this further convinced him that natural selection was not an adequate explanation for the evolution of the highly intelligent, moral nature of humankind. Nevertheless, Wallace had grounded his arguments on the insufficiency of natural selection independent of any spiritualist claims. Darwin recognized that Wallace had articulated in a forceful and clear manner many objections that other people also had raised and Darwin needed to address them. He was not always successful. For instance, he tried to claim that hairlessness might be useful, especially if humans had originated in tropical Africa. Hair was harmful because it hosted ticks and other infestations. But he eventually agreed that it was hard to make a strong case for the utility of hairlessness. Although Wallace now claimed that primitive humans did not need a big brain to survive, previously he had written that the "savage" faced great mental challenges of immense complexity and difficulty in his daily life. Wallace thought, "The intellectual labor of a good hunter or warrior considerably exceeds that of an ordinary Englishman." Thus, it was possible to make a utilitarian argument in favor of developing a large brain.

Clearly, the most difficult trait to account for by natural selection and the one that was consistently cited as what separated humans from the rest of the animal world was the development of a moral sense. Even if selection were capable of producing the beginnings of human reason and the moral sentiment, intensified social and sympathetic feelings would prevent the beneficial culling of the mentally and morally inferior. Therefore, natural selection would become disengaged. Wallace had recognized this problem. Although he argued that cooperation could be adaptive, the complex moral behavior of humans could not just be reduced to cooperation. This was one of the reasons why he argued that there must be some higher guiding power. Even before Wallace had published his views on the insufficiency of natural selection, these sorts of issues had been voiced by many others and were not trivial. Who better to take on the enemies of evolution than Darwin's self-proclaimed bulldog—Thomas Huxley. Huxley correctly perceived that the most controversial aspect of Darwin's theory concerned human ancestry.

In 1863, Huxley published *Man's Place in Nature*, one of the most articulate and eloquent explanations ever written of why human beings were no exception to evolutionary theory. He pointed out that the comparative anatomy of primates provided powerful evidence in favor of Darwin's theory (see figure 5.1). The differences among nonhuman primates were often greater than between humans and some of the higher primates. If Darwin's hypothesis explained the common ancestry of nonhuman primates, then it followed that humans also shared a common ancestor with them. Wallace certainly did not disagree with such an analysis.

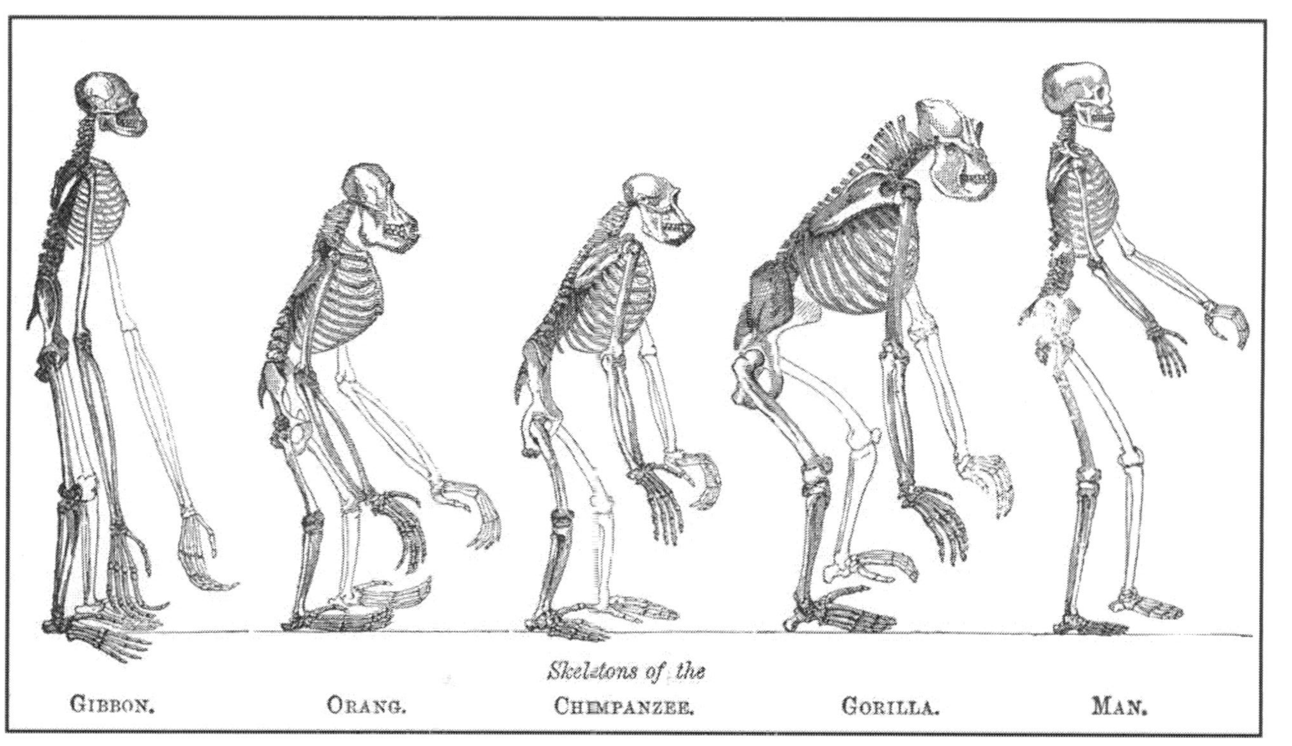

FIGURE 5.1. Comparison of Primate Skeletons. From Thomas Huxley, *Man's Place in Nature* (1863). Published in Sherrie Lynne Lyons, *Thomas Henry Huxley: The Evolution of a Scientist* (New York: Prometheus Books, 1999). Reprinted by permission of the publisher.

His two-stage theory of human evolution specifically explained such similarities. But this was not why Wallace and many others found Darwin's theory of human evolution incomplete. Huxley knew that the controversy surrounding human origins would not be settled on the grounds of comparative anatomy. Like Darwin, Huxley appealed to the Victorians' love of their pets to demonstrate the unity of humans with the animal world. "The dog, the cat . . . return love for our love and hatred for our hatred. They are capable of shame and sorrow." Anyone who had observed their actions recognized that they were capable of rational thought.[35] Thus, Huxley claimed a psychical as well as physical unity existed between man and beast.

Nevertheless, Huxley recognized that whether they were religiously devout or not, most Victorians would still be appalled at the inevitable conclusions to be drawn from his analysis. They would claim "the belief in the unity of origin of man and brutes involves the brutalization and degradation of the former."[36] But, Huxley questioned if that were really so. In a passionate entreaty, he claimed that human dignity did not depend on our physical characteristics or our origins. "Is it . . . true, that the Poet, or the Philosopher, or the Artist . . . is degraded from his high estate by the undoubted historical probability, not to say certainty, that he is the direct descendant of some naked and bestial savage. . . . Is mother-love vile because a hen shows it, or fidelity base because dogs possess it?"[37] It did not matter whether man's origin was distinct from all other animals or whether he was the result of modification from another mammal. Human dignity was not inherited, but rather to be won by each of us consciously seeking good and avoiding evil.

Furthermore, a vast chasm separated civilized humans from the "brute." Even if humans came *from* the brutes, they were not *of* them. Thus, Huxley's strategy to convince people of the truth of Darwinian theory was to claim that the highly charged issues concerning humanity's morals and ethics, questions of good and evil, were not relevant to the question of human origins. This is not to say that Huxley was uninterested in such questions, but he did not think the doctrine of evolution could give us an ethics to live by. "As the immoral sentiments have no less been evolved, there is so far, as much natural sanction for the one as the other. The thief and the murderer follow nature just as much as the philanthropist."[38] While evolution could provide an explanation for how the good and the evil tendencies of humankind may have come about, it did not explain why what we call good is preferable to what we call evil.[39] However, this did not mean that evolution was an inadequate or incomplete account of how humans came into existence.

Wallace disagreed. The essence of human beings was that they were spiritual beings and this is what provided the basis for ethical and moral behavior. Natural selection was inadequate to explain the evolution of humankind's moral and spiritual nature. Darwin may have tried to avoid dealing with human evolution, but this was the question that was on everyone's minds—not just Wallace's. Wallace's defection was a serious blow, but Lyell had also claimed that humans were the exception to the general evolutionary process. Brilliant as *Man's Place in Nature* was, Huxley did not address the problem of the evolution of mental and moral characteristics. Darwin could no longer avoid the problem of man. In 1871 in *The Descent of Man,* Darwin presented a theory of how the moral sense could have evolved.[40] In addition to his argument based on utility about the limits of natural selection, Wallace had used evidence for the existence of a spirit world to bolster his claim that some other force was guiding the evolution of humans toward higher and more ethical beings. Darwin needed to show that such evolution could occur without recourse to spiritual agents. Many if not most of Darwin's ideas Wallace fully supported. However, as we will see, Wallace and Darwin reacted quite differently to their encounters with indigenous people and this would be critical to their differing views concerning human evolution.

Just as Huxley had done in *Man's Place in Nature*, Darwin drew on comparative anatomy and embryology to demonstrate how similar humans were with apes. Moreover, humans even suffer from many of the same diseases. In *The Origin*, he had argued that homologous anatomical structures in humans and animals suggested they shared a common ancestor. He continued this same line of reasoning in *The Descent*, but in addition he focused on homologous mental structures. As was pointed out in chapter 3, basic emotions such as courage, fear, affection, shame, and fundamental mental faculties such as imitation, imagination, and reason were possessed by animals as well as by humans. The grief female monkeys expressed for the loss of their offspring, the curiosity of young apes, the jealousy and shame of dogs, and the reasoning abilities of higher animals all illustrated our shared intellectual and emotional heritage with the lower animals. Darwin used a variety of evidence to argue that no insuperable barrier existed that separated humans from lower animals.

Darwin maintained that even traits that most people considered unique to humans such as the ability to use tools, language, and an aesthetic sense existed to some degree in the lower animals. Apes manipulated sticks and pebbles for a variety of purposes and, while animals may not have symbolic language, they certainly were able to communicate information, some of which was quite complex. Women decorated their hats with brightly colored feathers just as male birds elaborately displayed their plumes and splendid colors before the female birds. Many similar examples existed in the natural theology literature to support the argument from design. But Darwin told a very different story. A lioness fighting till death to protect her young was not an example of the hand of God, but was used by Darwin to show how the moral sense could have evolved. Specifically, Darwin developed a theory of conscience.

Darwin made an assumption that was crucial to his theory, and one that most biologists and psychologists today depend on in trying to understand animal behavior, in spite of the risks associated with it. How does one justify attributing different mental states to other animals? Darwin reasoned that since the brain and nervous system of animals were similar to ours, then if the animal exhibited a particular behavior that had a human counterpart, he could assume that the underlying psychic state was similar as well. If I ask my son to bring me *The Origin of Species* and he does, I assume that he understood me. If I ask a dog to fetch the paper and he does, I assume at least at some level he understood me. When I pet my cat to show her affection and she likewise comes up to me and rubs and nuzzles me, I assume that she is showing me affection as well. How can we really know what the underlying mental state of an animal is when we don't even know what our fellow human beings are thinking most of the time? Perhaps we can't, but we make judgments all the time about the mental states of both people and animals based on their behavior. Mistakes are often made, but we also have learned a great deal about both animal and human mental life using this principle of psychological attribution. Darwin depended on this principle to build his case for the evolution of a moral sense.

Darwin's theory of conscience consisted of four overlapping stages. First, organisms developed a social instinct causing them to take pleasure in the company of others and bond together closely related and associated individuals into society. In the second stage, animals evolved sufficient intellect to recall instances when the social instincts went unsatisfied in order to satisfy stronger urges such as hunger or sexual drive. The development of language in the third stage enabled early humans to become sensitized to mutual needs and to codify principles of their behavior. Finally, habit would come to shape the conduct of individuals so even concerning small matters, acting in light of the wishes of the community would become in a sense second nature.

As was typical of the literature on animal behavior at the time, Darwin told animal stories to illustrate the many kinds of social behavior that animals exhibited. An old baboon heroically rescuing an infant attacked by dogs, rabbits stomping their feet in warning, old crows feeding their companions who were blind, were all examples of the social instincts in action. Although it was difficult to judge whether animals have any feeling for each other's sufferings, the many illustrations of helping and protective behavior from a variety of species suggested animals do express sympathy in some sort of rudimentary form.

Turning to human behavior, Darwin differed from the prevailing sensationalist view in psychology and this difference was crucial to his theory of conscience. Most British moral theorists and psychologists argued that sympathy was a learned behavior in response to pleasure and pain. For example, I feel sympathy for a person who cut her finger because I know what it feels like myself. But Darwin had a different analysis. Our sympathetic response to another was not a learned association, but rather was instinctive. Since the social and altruistic responses of animals were instinctive, why shouldn't they be so in us? However, there was an important difference between our behavior and that of animals. Humans were not limited to fixed patterns of behavior. Evidence from the behavior of the lower animals showed that humans had an instinctive urge toward sympathy, but how that sympathetic urge manifested itself was guided by reason. By claiming that social and altruistic behavior was instinctive, Darwin thought it was highly probable that an animal that had well-developed social instincts would inevitably acquire a moral sense or conscience as soon as its intellectual powers approached that of humans.

Thus social instincts alone were not enough to form what would be considered conscience or a moral sense. The intellect played a crucial role in the development of the moral sense, and the aspect of the intellect that was absolutely essential was memory. Memory allowed one to compare an unsatisfied social instinct with a more powerful urge such as hunger, fear, or sex. After the stronger urge had been satisfied, one would recall the unsatisfied social urge leading to a feeling of dissatisfaction. Eventually, one would remember in advance and reflect. In highly developed organisms such as ourselves, often these more basic urges would be sacrificed to satisfy the well-developed social instincts, such as when one gives up one's own life to save a drowning child.

For Darwin an evolved intellect played two important roles. First, reason and experience would guide conduct that had been stimulated by the social instincts. Although no specific instincts existed to tell one how to aid a fellow human being, the impulse to aid *was* instinctive. By living in a social setting and observing the actions of members of the community, one learned how to provide aid. This would become much more routine once speech had evolved in early human groups and be reinforced by the language of praise and blame. Second, a sufficiently evolved intellect allowed one to compare past and future actions or motives, approving or disapproving of them. Thus, according to Darwin, conscience and moral obligation ultimately was derived from the persistent social instincts. "The imperious word *ought* seems merely to imply the consciousness of the existence of a persistent instinct...serving him as a guide, though liable to be disobeyed. We hardly use the word *ought* in a metaphorical sense when we say hounds ought to hunt, pointers to point, and retrievers to retrieve their game. If they fail thus to act, they fail in their duty and act wrongly."[41] Darwin did not deny that a huge difference between animals and humans existed. Indeed, he did not consider animals moral beings. Retrievers could not be considered fully moral beings because they did not have the ability to reflect on their behavior and reestablish a suppressed social instinct. Nevertheless, he maintained the difference was a matter of degree only, not of kind.

Darwin's theory of conscience was not just based on utility, but depended on his claim that a continuum existed between humans and animals. He even suggested that the beginnings of religiosity could be found in the lower animals. His analysis of how the religious sentiment might have evolved was exceedingly shrewd. It was also extremely racist. He claimed that there was no evidence that indicated aboriginal man believed in the existence of an Omnipotent God. He wanted to distance the religious belief of aborigines from that of cultured society, claiming that their beliefs were totally distinct from each other. By making this distinction it meant that theologically minded Victorians could dismiss as superstition the natives belief in spirits and ghosts. Nevertheless, Victorians recognized that the "savage's" feelings were of a genuinely religious nature. Many native people believed that spirits animated natural objects. Wallace, living among these people, experienced their religious rituals that were permeated by the presence of spirits, and this undoubtedly made him more open to the spiritualist phenomena occurring in Victorian drawing rooms. However, Darwin had a very different view that drew in part on the work of anthropologist E. B. Tylor. Tylor had suggested that dreams gave rise to the notion of spirits because savages did not make clear distinctions between objective and subjective impressions. However, until the faculties of imagination, curiosity, and reason were sufficiently well developed in early humans, Darwin did not think their dreams would have led them to believe in spirits any more than in the case of a dog. He, however, then brazenly drew a parallel between the native "who imagined that natural objects and agencies were animated by spiritual or living essences" and the behavior of his dog which barked and growled at a parasol blown by the wind. He suggested the dog must have reasoned that its flight indicated the presence of an invisible agent.[42] The feeling of religious devotion of the savage, a combination of "love, complete submission to an exalted and mysterious superior, a strong sense of dependence, fear, reverence, gratitude, and hope for the future" was quite complex and could only be experienced by someone with moderately high moral and intellectual development. Nevertheless, it bore a strong resemblance to the simpler emotional response displayed by a dog's worshipful devotion to its master. The well-to-do English hunting man with his devoted hunting dogs would easily relate to such an image. Darwin obviously did not think that his comparing the religious feelings of native people to the behavior of a dog would be offensive to his readership. Rather, he hoped that his comparison would accomplish two things. First, by distancing the religious practices of "savages" from those of the "civilized races" he hoped to prevent a storm of criticism from pious Victorians. Although the same comparison could be made between the behavior of highly devoted Christians and a dog's behavior, Darwin knew better than to point out such similarities. Second, such a comparison implied that the underlying emotions of religious belief were incipient even in a dog! This would force people to acknowledge that there was a chain of continuity from animal to apelike ancestor to primitive human ancestor to modern-day "savage" to cultured Englishman.

However, Wallace had a quite different analysis of life's progression to higher and higher states. He insisted that an unbridgeable gap existed when it came to humans' moral and spiritual nature. He did not find Darwin's explanation for the evolution of religious feelings convincing in part because it rested on a very different view of native people. As the *Beagle* cruised along the shores of Tierra del Fuego Darwin glimpsed for the first time "wild men" in their native habitat and he was truly shocked.

> I shall never forget how savage & wild one group was. . . . They were absolutely naked & with long streaming hair, springing from the ground & waving their arms around their heads, they sent forth most hideous yells. Their appearance was so strange that it was scarcely like that of earthly inhabitants.[43]

Darwin could not believe how wide the difference was between the "savage" and civilized man. He thought it was even greater than between wild and domesticated animals because humans were capable of greater improvement. Could these naked barbarians, their bodies coated with paint, and their often-unintelligible gestures really belong to the same species as the Europeans? Darwin was shocked and repulsed, but he did not for a moment think that the different races actually represented different species. Like Wallace, Darwin was opposed to slavery and he knew that the theory of polygenesis had been used as a quasi-scientific argument to justify slavery. All human beings, no matter how diverse they were today, ultimately were derived from the same common stock.

Aboard the *Beagle* were three Fuegians—Jemmy Button, York Minster, and Fuegia Basket—who had been living in London for the last four years, but were being returned to their home. Darwin was continually struck by many little traits of character that showed how similar their minds were to Europeans. In their short time in England, the three Fuegians had become quite Anglicized in both manner and dress: Jemmy in his fine London clothes and Fuegia in her English bonnet. They were a stark contrast to members of their own tribe. When Darwin compared Jemmy to his real cousins, he was amazed to think that Jemmy undoubtedly had behaved in just the same manner as the "miserable savages whom we first met here." Clearly, the Fuegians did not represent some separate species. But the contrast between the Fuegians on board ship and those in their native land made Darwin realize both the tremendous potential for change and improvement and yet how tenuous was the state of civilization. In less than four years Jemmy had advanced from a stage of savagery to someone who polished his shoes and made jokes. This amazing plasticity exhibited by human beings would be crucial to his later formulation of a theory of evolution. Cruelty and brutishness as well as kindness and sympathy were all part of the human condition. If the gap between the civilized and savage races could be bridged, which was greater than that between domestic and wild animals, this would provide the foundation for his assertion that no unbridgeable gap existed between animal and humankind.

While Darwin recognized that the Fuegians were one and the same species as the Englishman, their "savage" behavior in their natural state would be something he would ponder for many years to come. He would later think that the present-day primitive races provided a window into the past, exhibiting behavior that was undoubtedly quite similar to that of ancestral primitive races. This would suggest a chain of continuity from an apelike ancestor to primitive human ancestor to present-day humans. However, Wallace had a very different view of the natives with whom he lived.

Wallace immediately perceived that the natives were no different from Europeans. They did not require a sojourn to London like Jemmy to learn "civilized" behavior—and that was precisely the problem. The so-called higher faculties were already fully developed, in spite of the harsh environment that these people lived in. Furthermore, he believed that the unsavory behavior such as lying and stealing that was attributed to the natives in the Malay Archipelago was a result of these people coming into contact with the Europeans. Wallace could not abide the behaviors of the "civilized" races who, driven by greed, were creating a society where corruption and dishonesty were rampant. He became increasingly critical of the "survival of the fittest" mentality that dominated life in Britain. For the supposedly better-off, but poor man in England, life was often worse than for the Malays or Burmese before they had any modern inventions. Wallace also thought that many natives clearly had superior mental and moral qualities, and therefore colonial rule by the Europeans was an absurdity.[44]

Wallace had become quite pessimistic about the civilized nations and believed that it was the mediocre in regard to both morality and intelligence that were succeeding best in life and multiplying. Nevertheless, he remained convinced that the human species continued its moral and intellectual advancement. However, this development was not due to "survival of the fittest." Rather, the improvement was due to an inherent progressive power that provided absolute proof for the existence of higher beings.

Spiritualism provided Wallace with a practical morality that resonated with his egalitarian views, his sense of justice, and his unequivocal idealism and optimism regarding the future of the human race. However, it would be a mistake to think that Wallace became converted to spiritualism because its ethical and metaphysical underpinnings were a substitute for traditional Christian doctrine that he had found unsatisfactory. Wallace believed that our place in nature was "strictly due to the action of natural law." Therefore, he also was arguing that spiritualist phenomena should be included within the purview of legitimate scientific investigation.

Spiritualism and the Boundaries of Science

Just as Crookes's investigations into spiritualism cannot be considered separately from his more mainstream scientific researches, Wallace's involvement with spiritualism was part of a coherent worldview he espoused. He wanted to develop an evolutionary model that was much broader than just explaining change within the physical/biological realm. In doing so he was able to accommodate both natural selection and spiritualism. Although Wallace referred to his new views on man as "unscientific" and that a few years back he would have regarded them as "wild" and "uncalled for," he also maintained that his opinions on the subject had changed solely because he had observed a series of remarkable physical and mental phenomena. He was convinced that these demonstrated unequivocally the existence of forces and influences not yet recognized by science.[45] Thus, for Wallace a belief in spiritualism was not incompatible with science. Rather, he wanted the boundaries of science to be extended to include phenomena that could not be explained in strictly materialistic terms. He regretted that he had titled his essay "The Scientific Aspect of the Supernatural." "Supernatural" was a misleading term because "all phenomena, however extraordinary are really "natural" involving no alteration whatsoever in ordinary laws of nature."[46] Although he thought that the origin of life resulted from the complex arrangements of inorganic molecules, he also thought that transition from lifeless to life was part of an overall plan that had been guided by conscious beings that were outside of and independent of matter. If molecules lacked consciousness, how could a complex arrangement of them produce consciousness?

In contrast, Huxley claimed that consciousness was simply the result of molecular changes in protoplasm. In 1868, Huxley appeared before an

Edinburgh audience with a bottle of smelling salts, water, and various other common substances, and told them that he had before him all the basic ingredients of protoplasm or what he translated as the physical basis of life. Matter and life were inseparably connected. All living organisms shared a common unity of form—the cell, and all cells shared a similar chemical composition. Plants and animals appeared to be very different, yet no sharp dividing line existed between the simplest of these organisms. Even the distinction between living and nonliving matter lay in the arrangement of molecules. He went further and maintained that the expression of intellect, emotion, and freewill could ultimately be explained in terms of material changes.[47]

Not surprisingly, Huxley's lecture threw the religious community into an uproar and he was accused of being an atheist, positivist, materialist, and worse. But Huxley vehemently denied being a materialist, claiming that both matter and the spirit were both names for an unknown and hypothetical cause or condition of states of consciousness. Huxley believed that the fundamental doctrines of philosophical materialism, like those of spiritualism, lay outside the "limits of philosophical inquiry."

Huxley may have claimed that it didn't matter whether matter was regarded as a form of thought, or thought was regarded as a form of matter, since each statement had a certain relative truth. However, he preferred the terminology of materialism because it provided a method of inquiry into the phenomena of nature, including the nature of thought, by studying physical conditions that were accessible to us. Furthermore, by describing spiritualist terminology as "barren" and leading to a "confusion of ideas," he implied that spiritualistic metaphysics itself was barren. Materialism might not answer everything, but for Huxley, spiritualistic metaphysics answered nothing, even if it posed interesting questions. If not a materialist, then how did Huxley describe his own beliefs? He coined the word "agnostic" to define his own philosophy.

Typically agnosticism is placed on a religious spectrum somewhere between the absolute certainty of Christian belief and the total denial of the existence of God by atheists.[48] However, this was not the meaning as Huxley originally intended. Rather, agnosticism represented an epistemological claim about the limits to knowledge. Huxley maintained that our knowledge of reality was restricted to the world of phenomena as revealed by experience. While he had no a priori objections to the doctrine of immortality, he had no reason to believe in it. He also had no means of disproving it. It was not that Huxley thought the question of immortality unimportant. Rather, he did not think it was fruitful to try and study problems that at the present stage of human knowledge were unsolvable. For Huxley, the question of immortality lay outside the boundary of topics that could be investigated by the scientific method.

Wallace, along with other scientists such as Crookes insisted that phenomena that provided evidence for immortality had been demonstrated scientifically. However, they certainly were going against the scientific mainstream, which regarded such investigations as marginal at best. At the most fundamental level they differed from most of their colleagues on the question of evidence, which is one reason why spiritualism remained at the edges of scientific respectability. Wallace, to the end of his life insisted that he had witnessed phenomena "always under such conditions as to render any kind of collusion or imposture altogether out of the question."[49] But Huxley, along with the majority of scientists was not persuaded. It was not just that most mediums were in fact frauds. In addition, research on spiritualism and psychic phenomena came up against the same fundamental criticism. The phenomena were incapable of being demonstrated by standard methods of observation and controlled experimentation under predictable circumstances. Rather, they were hopelessly erratic and subjective. It was not possible to do what could be truly scientific studies of mediums because the proper conditions were not met in a séance. Instead, there was (1) usually an absence of light, (2) lack of control over conditions being allowed, (3) the diversion of attention by the medium, (4) the element of surprise on the part of the medium, (5) the possibility of concealment in the room, (6) the power of suggestion by the medium, and (7) the emotional expectancy on the part of the sitters. Critics, as philosopher Karl Popper was to clarify later on, insisted that scientific investigations had to be based on hypotheses that could be proven false. A theory about the sensory-motor apparatus of the nervous system that was being developed by psychophysiologists was capable of being falsified. However, speculations about the separate existence of the mind from body, like arguments for the reality of spirit forces could never really be falsified. Virtually all the fundamental questions that the Society for Psychical Research investigated as well as the major assumptions of spiritualists could not be definitely proven or falsified under any conceivable test. Because of this, the majority of scientists felt that it was not worth either their time or energy to seriously investigate psychic phenomena since such investigations could not lead to reasonably certain knowledge. Science did not provide an answer to all the mysteries of life, and perhaps never could. But most scientists argued that to introduce some occult or spiritual agency or force that was prior to and superior to all law was to part company with scientific practice.

Wallace disagreed and insisted that spiritualism could be studied scientifically, and in that respect he drew a clear distinction between spiritualism and religion. Nevertheless, his spiritualism as well as his science became nested within an overall theistic belief that increasingly dominated his thinking as he matured. *The World of Life: A Manifestation of Creative*

Power, Directive Mind and Ultimate Purpose (1910) summarizing his views on Darwinian evolution, was a vitalistic presentation of findings in embryology, heredity, and physiology.[50] For Wallace, modern spiritualism demonstrated that some higher being was guiding the process of organic evolution in the direction of higher beings. After biological death the spirit continued to evolve, ultimately resulting in a higher entirely spiritual form. Thus, spiritualism was a complement to the process of organic evolution. In this view natural selection became subservient to a belief in a deity that had become an essential component to a complete investigation of nature.

Wallace's views wonderfully illustrate the difficulty in determining what counts as science. His explanation for the complexity and diversity in the cosmos was highly speculative, but speculation is critical to good creative science. Nevertheless, we can also see in Wallace's work why spiritualist investigations remained at the margins of science, rather than becoming mainstream. First, many people, including people who were quite sympathetic to spiritualism, thought Wallace was quite gullible and uncritical in evaluating the evidence. Equally, if not more important, a small but increasingly influential group of scientists were successful in arguing that spiritualist phenomena were outside the boundary of what they defined as "legitimate" scientific knowledge. There may be aspects of the universe that are truly incomprehensible to humans or not capable of being investigated by the scientific method. Although this may be a true statement, it is not a fruitful one for furthering the advance of knowledge. Huxley disagreed with many aspects of the Darwinian hypothesis, but he also recognized that it was a very powerful idea that generated questions in virtually all areas of the life sciences that could be empirically tested. However, the same could not be said for many of the spiritualist claims. Wallace postulated that "the Infinite being, foreseeing and determining the broad outlines of a universe" would "impress a sufficient number of his highest angels to create by their will-power the primal universe of ether." He also postulated a body of "organizing spirits who were responsible for thought-transferences."[51] How would it be possible to absolutely disprove such a claim?

The scientific establishment certainly did not speak in one authoritative voice regarding the controversies surrounding spiritualism, and considerable negotiating went on in determining what counted as "scientific." However, the search for the common denominators of the universe, for the basic building blocks of nature, motivated spiritualists and psychic researchers just as it inspired the scientists who criticized those people. To find a fundamental theory, a framework that would unify every diverse particle and force in the cosmos was basically the same in both groups. The physicists emphasized the links between heat, electricity, and magnetism while the psychic researcher looked for connections between mind, spirit, and matter. As findings from paleontology, geology, comparative anatomy, and embryology suggesting a continuity and uniformity in nature began to make their way into the imaginations of the public, it seemed more and more likely that a connection did exist between the apes and us. Such a link brought a sense of urgency to answer the age-old question of "man's place in the universe."

Scientists, along with spiritualists and psychical researchers, as members of Victorian society, shared goals that were central to the time in which they lived. They were reacting to what they perceived as the crass materialism that the new findings in science seemed to be moving toward, but at the same time they felt the need to bring religion more in line with the teaching of modern science. Many spiritualists and not a few scientists had a vision of a "new science" that would unify spirit and matter, mind and body. If such a vision could be achieved, it would reconcile the growing gulf between science and faith. Thus, these people took on the most critical issues of science, philosophy, and religion. Their work was not ridiculous, but was an attempt to come to terms with a changing world.

I have argued that the investigation of sea serpents, phrenology, and spiritualism when examined in their historical context do not deserve the pejorative label of pseudoscience. Nevertheless, they never became part of the scientific mainstream. While a variety of factors contributed to their marginalization, ultimately they remained on the fringes of scientific respectability because their research methods and evaluation of evidence did not meet the standards of sound scientific practice as defined by the emerging class of professional scientists. The next chapter explores some specific aspects of evolutionary theory that were problematic for Thomas Huxley and in doing so elucidates some of the reasons evolution had a different fate than the previous case studies.

6

Thatige Skepsis

Thomas Huxley and Evolutionary Theory

> Darwin's position might . . . have been even stronger than it is if he had not embarrassed himself with the aphorism "Natura non facit saltum" which turns up so often in his pages. We believe . . . that Nature does make jumps now and then and a recognition of that fact is of no small import in disposing of many minor objections to the doctrine of transmutation.
>
> Is it satisfactorily proved, in fact, that species may be originated by natural selection? that there is such a thing as natural selection?
>
> —Thomas Huxley, "The Origin of Species"

I HAVE SUGGESTED THAT in some ways Darwin's theory of evolution can be considered a marginal science in the Victorian period, sharing many similarities to phrenology in how it was regarded and by the fact that it was not incorporated into university curriculums. Like phrenology, evolution's early advocates pre-Darwin were found among the dissidents, in the secular anatomy schools and the radical nonconformist medical schools, while Cambridge, Oxford and Edinburgh universities were bastions of natural theology.[1] In Thomas Huxley's ceaseless quest for the professionalization of science, he led the charge to break the grip the Oxbridge natural theologians held over natural history. In a relatively short time evolution moved from the edges of scientific respectability to the scientific mainstream, and natural theology became marginalized as professional science came to be identified with people such as Huxley, Faraday and Tyndall.

As the previous chapters have demonstrated, Darwin's theory challenged fundamental concepts concerning "man's place in nature." While some religious thinkers welcomed evolutionary ideas, nevertheless, Darwinian theory contradicted many tenets of traditional Christian belief. Thomas Huxley quite correctly perceived that many of the scientific objections were at their root religiously based and claimed that "he was prepared to go to the stake if necessary" against the enemies of evolution. He proselytized the virtues of evolutionary theory wherever he went, using the theory to promote his larger political agenda of making science rather than the church the source of moral authority and power in society.

Although the conflict between science and religion was a major part of the debates surrounding evolutionary theory, and one in which Huxley played a leading role, it is not the focus of this chapter. Instead, Huxley's skepticism regarding particular aspects of Darwin's theory will be used to explore some of the scientific objections to evolution. In his famous *Times* review of *The Origin*, Huxley maintained that although Darwin's theory explained a great deal about the natural world, he preferred to adopt Goethe's aphorism "*Thatige Skepsis*" or active doubt in evaluating it. Huxley's doubts provide a window to view the emerging structure of what becomes defined as good scientific practice and suggest why unlike the other case studies that remained marginalized, evolutionary theory emerged as one of the most powerful and unifying theories for the life sciences. Nevertheless, we will also see that defining what was a "fact," a "hypothesis," a "theory," and what counts as evidence, was problematic and was negotiated.

Even before Huxley appointed himself as "Darwin's bulldog" Victorians had been exposed to evolutionary ideas in a variety of different contexts, as we have seen from previous chapters. Darwin may have wanted to distance himself from phrenology, but it played a crucial role in the acceptance of evolution. Gall advocated treating the mind in biological terms and his comparative approach showed that the difference between animals and humans was in degree only. Such ideas received enormous exposure with the phenomenal success of the *Vestiges of the Natural History of Creation* with its chapter on the mental constitution of animals. Furthermore, the *Vestiges* may have been full of references to the Creator and natural law, but there was no doubt that this was an explicitly evolutionary account of the history of life.

Although the *Vestiges* enjoyed enormous popular success, most men of science were highly critical of it. Huxley was particularly venomous in his review. Opening with a quote from *Macbeth*, "'Time was that when the brains were out, the man would die,'" Huxley did not understand why the

Vestiges had not shared a similar fate.[2] Describing the book as "a mass of pretentious nonsense" and a "notorious work of fiction," he was appalled that a tenth edition was being printed. Far more influential on Huxley and other men of science was Charles Lyell's *Principles of Geology*. Five editions of the *Principles* appeared in the 1830s and four more were published before 1859. Although Lyell had argued against transmutation in the *Principles*, for some people it had the opposite effect. He presented Lamarck's argument in favor of the inheritance of acquired characteristics so well that Herbert Spencer wrote, "My reading of Lyell, one of whose chapters was devoted to a refutation of Lamarck's views, had the effect of giving me a decided leaning to them."[3] More importantly, the *Principles* presented a totally naturalistic account of the history of life. As Huxley later wrote,

> I cannot but believe that Lyell for others as for myself was the chief agent in smoothing the road for Darwin. For consistent Uniformitarianism postulates evolution as much in the organic as in the inorganic world. The origin of a new species by other than ordinary agencies would be a vastly greater "catastrophe" than any of those which Lyell successfully eliminated from sober geological speculation.[4]

Furthermore, as historian Michael Bartholomew succinctly wrote, "Lyell's presentation of the species problem, as it stood in 1832 is in effect, if not intention, far less damning of the idea of transmutation than the rejection of Lamarck might imply."[5]

In addition, Darwin's theory was being evaluated in a society in the throes of rapid industrialization, with its laissez-faire economics and unbridled competition. In spite of the harshness of this world, many people agreed with Herbert Spencer that such ruthless competition did weed out the less fit, and thus led to the overall improvement of society. It was Spencer who coined the phrase "survival of the fittest" that Darwin would later adopt to describe the struggle for existence in the natural world and his mechanism of natural selection. Spencer's life work beginning with *Social Statics* (1851) and elaborated in further writings including *First Principles* (1862), attempted a grand evolutionary synthesis of both the inorganic and organic world. His general law of evolution was a process of spontaneous ordering and self-organization that led to over all progress. Darwin also believed that the struggle for existence yielded benefits, writing "that the war of nature is not incessant, . . . that death is generally prompt, and that the vigorous, the healthy, and the happy survive and multiply." Furthermore, natural selection was scrutinizing each variation, "rejecting that which is bad, preserving and adding all that is good; silently and insensibly working, whenever and wherever opportunity offers, at the improvement of each organic being in relation to its organic and inorganic conditions of life."[6] Natural selection acted slowly, but it resulted in greater adaptation and progress. Although it acted on individuals, this led to overall improvement to the species as a whole. Such ideas resonated with the Victorians sense that hard work and struggle would result in the betterment of one's self and in turn would reap benefits for society as a whole. However, it would be a mistake to conclude that Darwin merely took the prevailing ideas from social theory and grafted them onto the biological world.

The Origin of Species in addition to advocating a theory of evolution contained a vast compendium of facts: facts of comparative embryology and anatomy, the latest findings in geology, paleontology, and biogeography. Darwin wrote to animal and plant breeders as well as carried out numerous experiments on his own. He wove all this information into a compelling tale of how species slowly change through time by his mechanism of natural selection. Nevertheless, facts must always be interpreted. Huxley, in spite of being Darwin's foremost advocate was skeptical of Darwin's two basic tenets, gradualism and natural selection. How and why Huxley was able to be Darwin's leading promoter in spite of these objections highlights some critical criteria that resulted in Darwinian theory being accepted into the canon of legitimate science.

Rebel with a Cause

From rather modest beginnings, Thomas Huxley rose to become perhaps the most prominent Victorian scientist. This would not have been possible a century earlier. But it also meant that there was a pressure to achieve far beyond the expectations of previous generations. Huxley's life in many ways epitomized both the opportunities and stresses of Victorian life. He was born in 1825 in the country village of Ealing, the youngest of seven children of George and Rachel Withers Huxley. His childhood was for the most part uneventful. Huxley's schooling was irregular and most of what he learned was self-taught although his father was the master of a large semipublic school. Even as a young boy he had interests in a staggering array of subjects. From James Hutton, he learned about geology. From Sir William Hamilton's "The Philosophy of the Unconditioned" he embraced the skepticism that typified his mature thought. From Thomas Carlyle he developed sympathy for the poor that was later reinforced by his exposure to the squalor and poverty he saw in the East End of London. He taught himself German to read Goethe and Kant in the original. This would serve him well in later years, allowing him to become acquainted with the tremendous biological advances being made in Germany that few English men of science were able to follow. He also had an intense interest in metaphysical speculation, engaging everyone he could find on questions such as the nature of the soul and how it differed from matter.

Huxley began studying medicine at quite a young age and received a scholarship to the medical school attached to Charing Cross Hospital. Except for physiology, most of the medical curriculum bored him and he never practiced as a physician. Like many others who made their mark in the natural sciences, Huxley took a voyage around the world as the assistant surgeon on the HMS *Rattlesnake* from 1846 to 1850. This resulted in some of his most important scientific work. Huxley's life was one of incessant activity. He lectured at the School of Mines and was also a professor at the Royal College of Surgeons. Evenings were often spent speaking before working men or learned societies. He was president of numerous societies including the British Association for Advancement of Science, the Geological Society and the Ethnological Society. He played a crucial role in educational reform, particularly in advocating the teaching of laboratory sciences in the schools. Actively engaged in research in physiology, morphology, and paleontology, Huxley worked tirelessly to see basic biological research recognized as legitimate in its own right, separate from the medical school curriculum.

In spite of Huxley's personal success and his many scientific achievements, this did not translate into financial success. Much of Huxley's income was from the writing of popular essays and books. He, nevertheless, managed to write several hundred scientific monographs. Eventually Huxley's frenzied life of appointments, speaking, teaching, writing, and research finally took its toll, and in 1873 he suffered a complete breakdown. It was only through the generosity of his friends including Darwin who collected £2,100 that he was able to go abroad for a complete rest and recovery. Unlike Darwin and Lyell who were independently wealthy, Huxley had additional very personal and practical reasons for wanting to see science become fully professionalized.

Huxley's interests were not confined to science. He loved music, art and literature and was a member of two famous London clubs: the X Club and the Metaphysical Society. The Metaphysical Society's members represented the intellectual elite of London society and provided the perfect forum for Huxley to present his views on religion, the nature of knowledge and the great theological and metaphysical questions that have always occupied human thought. It was Huxley's membership in the Metaphysical Society that caused him to coin the word "agnostic" to describe his own belief system and to distinguish it from other -isms, such as positivism, materialism, atheism, and even empiricism. It was meant to be antithetic to the "gnostics" of church history who claimed to know so much about the very things of which he was ignorant. As mentioned previously, typically agnosticism is regarded as a belief midway between total certainty that God exists and total denial. However, this was not Huxley's original meaning. Rather, agnosticism represented an epistemological claim about the limits to knowledge. Building on the Kantian principle that the human mind had inherent limitations and further elaborated by Hume, Huxley maintained that our knowledge of reality was restricted to the world of phenomena as revealed by experience. It also became the cornerstone in defining what constitutes the practice of science.

Huxley's early work from the *Rattlesnake* voyage established his reputation within the scientific community, but it was his defense of Darwinism that brought him into the public spotlight. His famous encounter with Bishop Wilberforce at the 1860 meeting of the BAAS in which he said he would rather have an ape in his ancestry than be descended from someone like Wilberforce, was an important milestone in his career. Not only did it ensure that Darwin's theory received a fair hearing, but it let the public know that that he was a force to be reckoned with in the world of science and religion. However, Huxley's advocacy did not prevent him from voicing what he regarded as weaknesses in Darwin theory. Research in both embryology and paleontology suggested that organisms were grouped into distinct types with no transitional organisms between them. Such findings were problematic for a theory that claimed that one species gradually evolved into a different one. Instead, for Huxley evolution by saltation or jumps seemed to better fit the facts of development and the fossil record. Such a view also suggested that natural selection might be able to cause the formation of well-marked varieties, but lacked the power to create true species. Huxley eventually changed his mind about gradualism, but continued to doubt the efficacy of natural selection to cause speciation.

The Concept of Type

Long before Darwin presented his ideas to the world, natural philosophers had been searching for a classificatory scheme that reflected the natural groupings of organisms based on the similarity of structure. Huxley believed such a natural system of classification was possible because organisms could be grouped into distinct types. This concept of type was ubiquitous in the nineteenth century with the *Naturphilosophen* its most conspicuous advocates. Huxley had no use for the metaphysical claims of the Naturphilosophen. However, in Karl von Baer's research on development he found a usage of type that he would adopt in his own work. Von Baer's work demonstrated that the many varieties of animal forms were in actuality based on only four plans or types derived by simple laws of modification. The concept of type underlay von Baer's famous laws of development: (1) that the more general characters of a large group of animals appear earlier in their embryos than the more special characters; (2) from the most general forms the less general are developed, until finally the most special arise; (3) every embryo of a given animal form, instead of passing through the other forms, rather becomes separated from them; and (4) fundamentally, therefore, the embryo of a higher animal form is never identical to any other form, but only to its embryo.[7] Following von Baer's methodology of tracing the early stages of embryonic development and using the type concept as an organizing principle, Huxley demonstrated the interrelateness of a widely disparate group of organisms. However, like von Baer, Huxley argued that the four types were distinct, with no transition between them.

Cuvier also found that the animal kingdom could be grouped into four distinct types or *embranchements*: vertebrata, mollusca, articulata, and radiata. But rather than embryology, Cuvier found his four types in the comparative anatomy of adult organisms. The embranchement was not a taxonomic category that had been invented for the purposes of classification. Rather, organisms shared basic plans because they carried out similar interrelated functions. Every organ in the body was related and dependent on every other organ in order to maintain the functional integrity of the organism as a whole. It was not possible to modify one organ without modifying others and still maintain harmony. This principle explained the gaps found between various groups, particularly between the embranchements. He did not deny individual variation within a species, but he regarded such variations as restricted by the type of animal it was.

In the 1840s and 1850s, Huxley followed Cuvier and von Baer in arguing not only that distinct types existed, but also that they were fixed. Although von Baer allowed for a limited amount of evolution within groups, there could be no transformation of one group to another. Cuvier and von Baer did not argue for the fixity of type from idealistic principles or religious grounds. Rather, they made their case from empirical evidence, but they had enough theoretical assumptions to determine what empirical evidence would count. In any case, Huxley found that evidence quite convincing. However, Darwin used the evidence from embryology and paleontology to tell a very different story and forced Huxley to reevaluate his position.

Huxley's research had been guided by a fundamental problem: to define the doctrine of animal form. Classification was crucial to this enterprise because its purpose was to consolidate knowledge of isolated individual varieties of form into the fewest possible and most general propositions. Key to accomplishing such a task was to develop criteria for determining which resemblances between organisms were important and were indicative of a close relationship and which ones were superficial. Huxley decided that important resemblances were ones that were indicative of other resemblances and he defined them as affinities while all other resemblances were analogies. The distinction he used was quite similar to the one that Richard Owen made between homology and analogy and which became a standard tool for classification. Owen defined an analog as "a part or organ in one animal which has the same function as another part or organ in a different animal" and a homologue as "the same organ in different animals under every variety of form and function."[8] Huxley decided that structural similarity, therefore, was not what determined whether two organs were homologous, but rather that the organs had a similar law of growth. By studying patterns of development von Baer had demonstrated how forms could be transformed into one another. Furthermore, if two forms were similar, even identical, they were merely analogous if they had come about by different processes of growth. Thus, von Baer's methodology of tracing the developmental processes provided a practical way of determining the relationships between various organisms, and in doing so suggested that a natural classificatory scheme would be one that reflected the laws of growth shaping the form of the organism.

From our present-day evolutionary vantage point, the distinction between affinity (or homology) and analogy seems to shout out for a classification system based on descent from a common ancestor. Owen's definition of homology has stood the test of time precisely because it was compatible with Darwin's theory of descent. However, there is no indication in the 1840s or early 1850s that Huxley went beyond the embryological history of an individual (ontogeny) to speculating on the historical development of an entire lineage (phylogeny). Rather, the researches of von Baer, Cuvier as well as his own indicated that there were distinct types that precluded the possibility of transmutation. It remained for Darwin to employ von Baer's ideas of type in advancing evolution.

The relationship of development to the fossil record was hotly debated in the controversy over evolution. Leading biologists in the 1840s and 1850s all recognized certain parallels between the sequence of the fossil record and the stages of embryological development. But unlike Darwin, they used the "laws of development" to argue explicitly against transmutation. They agreed that the history of the earth revealed a succession of organisms appearing and disappearing. However, they claimed that evidence for a lower form evolving into a higher one was virtually nonexistent. Instead, they argued that the divergence observed both in fetal development and the fossil sequence seemed to be part of a wider developmental plan.

Darwin also linked embryology to paleontology, but came to a very different conclusion maintaining, "Embryology is to me by far the strongest single class of facts in favour of change of forms."[9] He did not think that the similarities between fetal development and the fossil record were merely analogous. Rather, Darwin's ancient progenitors were the archetypes of extant animal species. He believed that in these ancient animals the adult form and the embryo were similar. Thus, even in modern highly developed species the embryo still had the general unmodified form of the archetype. For von Baer, these resemblances were the necessary consequence of development from a common starting point by a single process of increasing specialization. For Darwin, the pattern in the fossil record was the result of descent from a common ancestor with divergence and increasing specialization occurring over time. The variations from the general archetype were inherited, but they usually did not make their appearance until late in development, the embryonic stages remaining unchanged. For Darwin, the archetype was a real organism that had existed in the distant past. Far from thinking that the type-concept was contrary to evolutionary thinking, Darwin found it crucial in dealing with embryology and the fossil record.

Darwin's ideas made immediate sense to Huxley. While he had no use for Lamarckian evolution, upon reading *The Origin* he exclaimed, "How extremely stupid not to have thought of that!"[10] The archetype of the morphologists had been determined by tracing patterns of development to determine which structures were analogous and which were homologous, with Owen clearly articulating what those two terms meant. This distinction had been a crucial criterion for classification. Darwin realized that good naturalists in their search for affinities were trying to distinguish superficial similarities from more meaningful ones, although they did not realize that those patterns were the result of descent from a common ancestor. He utilized the same kinds of evidence that taxonomists had found useful for purposes of classification to support his theory of descent from a common ancestor. This aspect of Darwin's theory was quite quickly accepted because it provided the theoretical underpinnings to classification. Descent with modification told taxonomists why organisms could be arranged in certain ways, but the outlines of that arrangement were well in place pre-*Origin*.[11]

From Saltation to Gradualism

Research in development supported the idea of distinct types, but Huxley recognized that it also provided evidence for common descent. Darwin's theory, nevertheless, presented a dilemma for Huxley. How could one reconcile evolution with the concept of distinct types? Furthermore, while the fossil record demonstrated unequivocally that vastly different organisms have inhabited the earth during different periods of earth history, it did not show a graduated chain connecting one group of organisms to another. Instead, groups became extinct and new forms abruptly appeared with no evidence of transitional forms. If transmutation occurred, where were the transitional forms?

Although Lyell advocated slow gradual change for the inorganic world, he was a saltationalist in his thinking about the biological world. By some unknown, but naturalistic means the earth was successively repopulated with different organisms in response to changing conditions. Rather than gradually changing to adapt to a changing environment, organisms appeared all at once, perfectly adapted. Huxley was grappling with the concept of type, the lack of transitional forms, and the uniformitarianism of Lyell. How could he put all these ideas together in a way that would be compatible with the theory of transmutation? Huxley's solution to this quandary was saltation. Huxley thought that the laws that governed the modification of organic bodies might be similar to chemical laws. Thus, when one atom substituted for another in a chemical compound, each modification was a discrete step; there were no intermediate stages. This analogy had the added benefit because it suggests that saltative evolution need not imply some kind of miraculous intervention. Darwin's theory was a naturalistic explanation of species change, just as uniformitarianism was a naturalistic explanation for geological change. Saltation allowed Huxley to explain the gaps in the fossil record, accept evolution, and most important, maintain a belief in the concept of type.

Darwin had borrowed Lyell's idea of slow geological change and applied it to the organic world, ignoring Lyell's biological views. But Huxley agreed with Lyell. For Huxley, evolution by "jumps" represented a more accurate description of the history of life. He cautioned Darwin on the eve of the publication of *The Origin,* "You have loaded yourself with an unnecessary difficulty in adopting *natura non facit saltum* (nature does not make jumps) so unreservedly."[12] Darwin was well aware of the problems that the fossil record presented for his gradualist scheme. "Geology assuredly does not reveal any such finely graduated organic chain, and this perhaps is the most obvious and serious objection which can be urged against my theory."[13] He devoted two chapters in *The Origin* to the fossil record to convince his readers that while it couldn't be cited in support of his theory, at least it wasn't against it. He borrowed an analogy from Lyell who said that reading the geological record was like reading a book where most of the pages were missing and only a few words were on each page. In other words, the fossil record was woefully imperfect. Darwin argued that the gaps in the fossil record were due to the irregular process of fossilization, competitive exclusion, and migration into new areas. Whether an organism was actually preserved or not depended on so many different factors that the fossil record represented a quite incomplete chronicle of the history of life. Moreover, this was the reason that it was so rare to see transitional forms between many present-day groups. Although birds appeared to be totally distinct from other vertebrate animals, he believed that the ancestors of birds were connected to the ancestors of other vertebrate classes. This particular example would play a crucial role in Huxley abandoning his saltational views. It also illustrates the kind of evidence that contributed to evolutionary theory being accepted.

In 1861, a remarkable fossil was found in the Jurassic limestone at Solnhofen in Bavaria—what appeared to be a feathered bird. Named *Archaeopteryx*, it was classified it as a bird, but an extremely primitive one that in many ways resembled a more general vertebrate type, rather than a modern bird. Although Huxley already thought that birds represented an extremely modified and aberrant reptilian type, he did not immediately jump on the evolutionary implications of *Archaeopteryx*. Too busy defending the highly controversial implications of evolution for human origins, the significance of *Archaeopteryx* would have to wait.

With more fossils constantly being discovered, Huxley continued his research on fossils and classification. Comparative embryology and anatomy demonstrated many similarities between reptiles and birds. Since many reptiles differed little from the general vertebrate type while all birds deviated significantly, he assumed that the reptilian features were common to both classes. Nevertheless, there were also several features that indicated reptiles and birds represented distinct classes. Were there any connecting links between the two groups? In 1868 Huxley made an extremely provocative and prophetic suggestion: Dinosaurs were the antecedent form between reptiles and birds. Although *Archaeopteryx* had many reptilian features, it was still a bird and, therefore, it could not be considered to be a true connecting link between reptiles and birds. Perhaps the pterodactyls or flying reptiles were the transitional form. They had air cavities in their bones, and various bones including the scapula and broad sternum were wonderfully birdlike. However, there were also important differences. They were similar to birds in the ways that bats among mammals were similar, in other words they represented "a sort of reptilian bat." They did not represent a link between reptiles and birds any more than a bat was a link between mammals and birds.

A large number of small birdlike dinosaurs had recently been discovered (see figure 6.1). Huxley was confident that older rocks would reveal birds more reptilian than Archaeopteryx and reptiles more birdlike than these dinosaurs, eventually obliterating completely the gap that they still left between reptiles and birds. The structure of the pelvis in *Megalosaurus*, *Iguanodon*, and *Hypsilophodon* were intermediate between reptiles and birds. Since not all the dinosaurs showed these modifications, they were evidence in favor of gradualism because they showed a series of modifications, from bones that were very reptilian to ones that were virtually indistinguishable from birds. The final type of evidence that Huxley found especially compelling for the close relationship between birds and dinosaurs could be found in certain species of present-day birds. Particular bones in the dorking fowl if found in a fossil state would be indistinguishable from that of a dinosaur. Moreover, if it were somehow possible to enlarge and fossilize the hindquarters of a half hatched chicken, it would be classified as a dinosaur. The dinosaurs provided crucial evidence in support of gradual evolution because they represented transitional organisms across the boundaries of widely separated classes.

Ever the skeptic, however, Huxley did not think that the evidence unequivocally proved that birds had arisen through gradual modification of a dinosaurian form, but only that such a process *may* have taken place. For Huxley, truly demonstrative evidence would be to find a series of forms that progressed in a steplike manner showing the metamorphosis of a reptilian-like dinosaur to a birdlike dinosaur to a bird. While

FIGURE 6.1. FOSSIL OF *Compsognathus longipes*. This small bipedal dinosaur with birdlike limbs was one of the pieces of evidence that Huxley used to argue that dinosaurs were the connecting link between birds and reptiles. From Thomas Huxley, "On the Animals Which Are Most Nearly Intermediate between Birds and Reptiles," in *Scientific Memoirs of Thomas Henry Huxley*, 4 vols., ed. M. Foster and E. Ray Lancaster (London: Macmillan, 1898–1902), vol. 3. Published in Sherrie Lynne Lyons, *Thomas Henry Huxley: The Evolution of a Scientist* (New York: Prometheus Books, 1999). Reprinted by permission of the publisher.

numerous transitional forms existed, they were found in contemporaneous deposits, not in the exact order in which they should be if they really had formed a progression from reptile to bird. This type of evidence did not yet exist for the reptile-bird transition, but it did exist for other parts of the animal kingdom.

For Huxley, the most persuasive evidence in favor of gradual evolutionary change was a series of fossil horses discovered in North America. The horse lineage proceeded from a horse form that represented a generalized mammalian type to a form that was highly specialized for its mode of life. The general mammalian type had a distinct radius and ulna, both of which were movable, five toes, both front and back, and a complete tibia and fibula, that were also distinct separable bones. Was there a sequence of fossil horses which through time showed a reduction of the number of their toes, a reduction of the fibula, a more complete coalescence of the ulna with the radius, leading to the highly specialized form of the modern-day horse? The answer was yes. In the most recent Pliocene deposits of Europe were fossil horses that were virtually indistinguishable from modern-day horses. From the older Pliocene and Miocene deposits came fossils horses that were also extremely similar to modern ones, but with certain significant differences. One, named *Hipparion*, had attached to the long metacarpal bone two splintlike bones and attached to the end of them was a small toe with three joints of the same general character as the middle toe. They were very much smaller and must have been essentially nonfunctional, similar to the dewclaws (vestigial digits) found in many ruminant animals. Nevertheless, these lateral toes were fully developed in *Hipparion* while they were almost abortive in modern horses. The hind limb had a fibula that was essentially the same as the modern horse, but the forelimb had a distinct ulna. From the middle and older parts of the Miocene, a series of fossils had been found that were still clearly like the horse in their general organization. However, in these forms, all four feet had three distinct functional toes, not the dewclaws of *Hipparion*. The lower Miocene form *Mesohippus* had three toes in front and a large splint like rudiment representing the little finger. The radius and ulna were entire bones and the tibia and fibula were also distinct. *Orohippus*, from the lower part of the Eocene, had four complete toes on the front limb, three on the hind, a well-developed ulna, and a well-developed fibula. In addition to the skeleton, teeth were often an important characteristic used in classification. This series of fossil horses also showed changes in their tooth structure and arrangement, going from a generalized mammalian type to a more complicated pattern of canines and molars with accompanying structural changes in the incisors and molars that was adapted for the diet of the modern horse.

While there was no way to know whether these fossil species represented the exact line of modification leading to the modern horse, these forms represented a succession of the horse type with the oldest being closest to the general mammalian type. Huxley maintained this succession of forms were "exactly and precisely that which could have been predicted from a knowledge of the principles of evolution," that is, forms gradually would become better adapted to their environment, becoming more specialized over time.[14] In a lecture to the Americans on evolution, Huxley predicted that equine fossils from the Cretaceous would have four complete toes with a rudiment of the innermost toe in front and probably a rudiment of the fourth toe in the hind foot. Two months after he gave the lecture, such a form was discovered from the lowest Eocene deposits of the West.

Fossil evidence had been the primary reason why Huxley advocated saltative evolution. As the nature of the evidence changed so did his views. He now accepted that Darwin had been correct in claiming that the incompleteness of the fossil record was responsible for the lack of transitional forms. The sequence of fossil horses was not the only evidence in favor of gradual change. As the century progressed, more and more fossil "connecting links" continued to turn up. A series of fossils showing the transition from the ancient civets to modern hyenas were discovered as well as seventeen varieties of the genus *Cynodictis* that filled up the gap between the viverine animals and the bearlike dog *Amphicyon*. In 1859, it appeared that a very sharp and clear division existed between vertebrates and invertebrates, not only in their structure, but also more importantly in their development. Twenty years later, while the exact details of the connection between the two groups were still not known, investigations on the development of *Amphioxus* and of the *Tunicata* demonstrated that the differences that were supposed to constitute a barrier between the two were nonexistent. Another complete separation appeared to exist between flowering and flowerless plants. However, research on the modifications of the reproductive apparatus showed that the ferns and the mosses were gradually connected to the flowering plants. Even the absolute distinction between plants and animals was breaking down with the discovery of simple life-forms that had characteristics of both kingdoms.

By 1874 Huxley was convinced that "there is conclusive evidence that the present species of animals and plants have arisen by gradual modification of preexisting species."[15] Facts from many different disciplines overwhelmingly supported the idea that species had changed through a perfectly naturalistic process. Although many gaps remained, the fossil record was providing links between different classes and documenting serial modifications of particular lineages. Even more dramatic evidence in favor of the common ancestry of all organisms was the research in development and comparative anatomy. Twenty one years after the publication of *The Origin*, Huxley claimed that "evolution is no longer a speculation, but a statement of historical fact."[16] Yet Huxley remained skeptical his entire life of the ability of natural selection to create new species causing Darwin to refer to him as the "Objector General" on the matter.

Natural Selection and the Problem of Hybrid Sterility

Huxley had no doubt that natural selection played an important role in the history of life, providing an explanation for many apparent anomalies in the distribution of living beings in time and space. But was it sufficient for the creation of new species? *The Origin* was filled with endless examples of the variety that had been generated by artificial selection. Huxley agreed with Darwin that there appeared to be no limit to the amount of divergence that could be produced by artificial selection and that the races formed had a strong tendency to reproduce themselves. "If certain breeds of dogs, or of pigeons, or of horses, were known only in a fossil state, no naturalist would hesitate in regarding them as a distinct species." However, all these examples had been produced by human interference. "Without the breeder there would be no selection and without selection no race."[17] Darwin argued that in the wild it was Nature herself who took the place of the breeder. In the struggle for existence those individuals with variations that allowed them to compete most successfully survived in greater numbers, passing on those traits to their offspring, and through successive generations they eventually acquired the characters of a new species. While Huxley did not think that the evidence contradicted such an explanation, natural selection was still just a hypothesis, not a theory. It had not been demonstrated that selection whether artificial or natural had been the responsible agent for the creation of new species as opposed to well-marked varieties.

For Huxley, a hypothesis obtained the status of theory when it had overwhelming evidence in the form of facts to support it. However, as has been illustrated throughout this book, what constitutes a fact is often difficult to determine and to some extent is in the eye of the beholder. Huxley was naive in his confidence that the "facts of nature" were just "out there" waiting to be discovered. His doubts over natural selection were in part due to a different view from Darwin over what constituted proof of a hypothesis. He also had different criteria than Darwin for what characteristics were crucial in defining a species, and that difference revolved around the question of hybrid sterility. The "facts" did not just speak for themselves.

Groups that had the morphological character of species, that is, distinct and permanent races, had been produced many times both in the wild and by artificial breeding. However, Huxley argued that there was no evidence that any group of animals had by variation and selective breeding given rise to another group that was in the least degree infertile with the first. He drew a distinction between morphological and physiological species. Morphological species were not true species at all according to Huxley. Rather, they were varieties or races that only *looked* like distinct species. Breeders had been quite successful in creating such "species." Pigeon fanciers, by artificial selection had made distinct races such as the pouter, carrier, fantail, and tumbler. Huxley did not deny that if fossils of the fantail or pouter were found paleontologists would regard them as good and distinct morphological species. But they failed his criterion for physiological species in that the offspring between the races were perfectly fertile and they were descended from a common stock, the rock pigeon. On the other hand, a cross between a horse and an ass resulted in the sterile donkey. A horse and an ass were physiological species, but how had they arisen? For Huxley, good, true physiological species were by definition incapable of interbreeding or produced offspring that were sterile. He drew a clear distinction between varieties or races that could still interbreed and species that could not. Only when physiological species had been observed that had emerged from varieties in the wild or had been produced by means of artificial breeding would Huxley consider that natural selection had been definitively proven.

Darwin had a very different view of species claiming that "there is no essential distinction between species and varieties." Sterility was "not a specially acquired or endowed quality, but is incidental on other acquired differences."[18] Nevertheless, he took Huxley's objections quite seriously and enlisted the aid of various breeders as well as continued to do experiments himself in an effort to convince Huxley that the sterility problem was not insurmountable. He pointed out that in nature there were many peculiar cases of unequal reciprocity of fertility between two species. Some species were more fertile with foreign pollen than with their own.

Darwin recognized that if he could show that sterility was an acquired or *selected* character, natural selection would be on much firmer ground. As a result of his research on *Primula*, for a short time he thought this might be the case.[19] It was well known by breeders that cross-fertilization increased the vigor and fertility of offspring while self-fertilization diminished them. Thus, Darwin reasoned that it would be an advantage for plants to develop a means of preventing self-fertilization. In *Primula* this was exactly what had happened. The primrose and cowslip each existed in two distinct forms. In one form the style was long and the stamens were short, while in other form the style was short and the stamens long. The two forms were distinct with no gradation between them and were much too regular and constant to be due to mere variability. Darwin did crosses of all possible combinations and concluded that heterostyly appeared to be an adaptation to prevent self-fertilization. Having made the suggestion that self-sterility had a selective origin, Darwin tentatively asked whether cross-sterility might also have originated by selection. Numerous examples existed of partial sterility resulting from the crossing of varieties. If Darwin could demonstrate that sterility between varieties was being selected for to preserve the integrity of incipient species, this would answer Huxley's objections. He continued his research and enlisted the aid of a young horticulturist John Scott, who by selective breeding succeeded in creating varieties of *Primula* that when crossed produced offspring that were completely sterile. Huxley's criteria of creating a physiological species had finally been met.

Much to Darwin's dismay Huxley continued to bring up "his old line" about the hybrid sterility difficulty, in spite of being well aware of confusing and contradictory data on hybridization. He acknowledged that varieties of plants existed when crossed whose offspring were almost sterile, while in both animals and plants that had been regarded by naturalists as distinct species, when crossed produced normal offspring. Some plants gave the anomalous result that pollen of species A could successfully fertilize the ova of species B, but the reverse was not true—a cross between pollen from species B and ova from species A was ineffective. The sterility or fertility of crosses appeared to bear no relation to the structural similarities or differences of the members of any two groups. He also realized there were many practical difficulties in applying his hybridization test. Many wild animals wouldn't breed in captivity. The negative results of such crosses were meaningless. Since many plants were hermaphrodites, it was difficult to be positive that the plants when crossed had not been self-fertilized as well. In addition, experiments to determine the fertility of hybrids must be continued for a long time—particularly in the case of animals. Nevertheless, while selective breeding had produced structural divergences as great as those of species, it had not succeeded in producing equal physiological divergences. Until this physiological criterion of sterility had been met Huxley maintained that natural selection had not produced all the phenomena occurring in nature.

Darwin thought that Huxley was demanding a type of evidence that would be essentially impossible to find. In fact, some of his research indicated that speciation would *not* occur through the agency of artificial selection. He noted that in several cases two or three species had blended together and were now fertile. The dispute between the two men was further complicated because Darwin went back to his original position on the origin of hybrid sterility. Although Darwin agreed with Huxley that hybrid sterility was an important measure of speciation, the opposite argument could be made as well. The fact that offspring from crosses between different varieties were often partially sterile while crosses between individuals that botanists claimed were distinct species sometimes resulted in fertile offspring was evidence that hybrid sterility was *not* a good criterion for defining species. If this was the case, there was no reason that sterility factors should be selected for. Further research convinced Darwin that his original view was correct. Sterility was an incidental result of differences in the reproductive systems of the parent-species. Huxley and Darwin were as far apart in their views as when they began their dispute.

Huxley never moved from his original position, but he also thought that since both the causes of variation and the laws of inheritance were still unknown, that an understanding of these phenomena would undoubtedly resolve this problem over hybrid sterility. Nevertheless, the disagreement between Darwin and Huxley was actually more fundamental than a dispute over the interpretation of experimental results. Huxley did not disagree with Darwin's analysis of hybrid sterility, but he had different criteria for what constituted proof of a hypothesis. He insisted that since true physiological species could not interbreed successfully, until such species had been produced either artificially or in the wild, the power of natural selection remained questionable. Since Darwin did not believe that sterility was being selected for, but merely a byproduct of other factors contributing to speciation, he did not regard the problem of hybrid sterility as undermining the power of natural selection to create new species. Instead, he cited a large amount of other evidence that supported natural selection as the primary mechanism for speciation.

What Counts as Proof?

Huxley's disagreement with Darwin raises a number of significant issues concerning the scientific process. How is evidence evaluated and weighted? When does the evidence become "enough" to claim that something is proven? In other words, when does something move from speculation to hypothesis to theory to fact? In the nineteenth century two main schools of thought competed for what was the "correct" scientific methodology. William Herschel maintained that we must argue from sense experience. A *vera causa* must be based on direct empirical evidence. However William Whewell claimed that we must argue *to* sense experience. His vera causa was based on the idea of consilience.[20] In *The Origin* Darwin made use of both kinds of arguments, but his case for natural selection relied on consilience. He argued not that there was direct experimental evidence for natural selection, but that it had wide explanatory power. Huxley acknowledged that fertility or infertility might be of little value as a test for speciation. Nevertheless, he maintained that any theory on the origin of species to be complete must give an account of how species became reproductively isolated and, therefore, infertile with each other if crossed. For Huxley, definitive proof of natural selection would have to be by empirical demonstration.

Darwin realized that Huxley was making basically an impossible demand on his theory. What was crucial for Darwin was not experimental demonstration of a hypothesis, but rather that the hypothesis could explain a great deal of observed phenomena including the geological succession of organic beings, their distribution in past and present times, and their mutual affinities and homologies. Darwin gathered a great deal of evidence in favor of his theory. He relied heavily on the analogy of artificial selection as evidence for his theory. However, it was an analogy, and as it came under increasing attack, primarily from Huxley, Darwin fell back on a more general claim for his theory of natural selection. If it explained large classes of facts, the theory deserved to be accepted.

Huxley's position regarding natural selection is somewhat of an enigma. Perhaps the explanation lies in his use of the term "physiological species." Physiology was just becoming established as a true *experimental* science and Huxley thought that experimental physiology would eventually provide direct proof for natural selection by doing the kind of experiments that would never be possible in paleontology. However, his demand for empirical proof for natural selection did not prevent him from recognizing the great theoretical power of the general theory of evolution. Huxley also drew heavily on the idea of consilience, continually asserting that the facts of paleontology, embryology, and taxonomy could best be explained by Darwin's theory. Although I claimed that two different types of arguments were used in support of scientific theories, the empirical vera causa of Herschel and the rationalist vera causa of Whewell, Huxley's defense of evolution points out that the distinction between the two was not absolute. Consilience would have no force in the absence of empirical evidence for the elements being brought together. Huxley's views also highlight both the similarities and differences between evolution and the other case studies in this book.

Like the other studies in this book, observations and "facts" were used as evidence for different positions. Huxley initially thought that the research of Cuvier, Von Baer, and his own work provided compelling evidence in favor of distinct types and the fixity of species. Darwin looked at the findings from comparative anatomy and embryology along with the fossil record and offered a very different interpretation. In addition, religious beliefs, issues of professionalization, the personal relationships between various scientists, and politics were also part and parcel of the debates over evolution. Although Huxley prided himself for his empiricism, these kinds of issues shaped his defense of evolution as well as influenced his interpretation of evidence. This is most clearly seen in his views on progression and the fossil record.

Huxley's desire to keep theological questions distinct from scientific ones underlies much of his scientific work and a great deal more of his popular writing. As was pointed out in previous chapters, one of the most heated debates in the nineteenth century concerned progression in the fossil record. By the 1830s most geologists thought that the stratigraphical record documented that the earth was several hundred thousand years old and more complex organisms appeared over time. However, Huxley agreed with Lyell's minority view point, maintaining that some types seemed to persist unchanged through vast periods of geological time. Both men, when they were president of the Geological Society of London used their presidential addresses to argue against progression. While a strong case could be made against progression in the 1850s, the fossil discoveries in the next decade began more forcefully to evince development. In 1863, even Lyell had acknowledged progression in the organic world with the exception of humans. Yet Huxley continued to argue against it. Undoubtedly Huxley's belief in the type concept played a role in his rejection of progression, but he still could have argued that distinct persistent types progressively appeared in the fossil record. However, Huxley associated the argument from design with progression. Many paleontologists, including William Buckland, Louis Agassiz, Adam Sedwick and Richard Owen were looking for a way to reconcile the findings of geology with Christianity and they believed that the progressive pattern of earth history was evidence for the argument from design. Ironically, as long as Huxley was unable to disassociate progression from the idea of design, he continued to ignore the increasing evidence in favor of progression.[21]

Huxley's obsession with keeping theology out of science was reflected in his scathing review of the *Vestiges*, which he later regretted. "The only review I ever have qualms of conscience about, on the grounds of needless savagery, is the one I wrote on the *Vestiges*."[22] Although *Vestiges* clearly presented an evolutionary account of the history of life, he objected to a scientific theory being used to bolster the idea of Divine Creation. He devoted most of his attention to the fundamental proposition of the book which was, quoting the author's own words: "'The Natural proposition of the 'Vestiges' is creation in the manner of law, that is the Creator working in a natural course or by natural means (Proofs lix).'"[23] Huxley paraphrased this definition as "Creation took place in an orderly manner, by the direct agency of the Deity" and in doing so attempted to reduce Chamber's entire book to nothing more than another treatise on natural theology. *Vestiges* described events of natural history, claiming the order of events was the result of natural law, which he then attributed to the action of a Creator. What were some of the natural laws the Vestigiarian cited? The primary one was that of progression. Huxley acknowledged that progression was a scientific proposition that could be accepted or rejected according to available evidence. He did not think the evidence supported progression. But even if "fully proved it would not be . . . an *explanation* of creation; such creation in the manner of natural law would . . . simply be an orderly miracle." Huxley believed "natural laws" were "nothing but an epitome of the observed history of the universe."[24] Thus, to claim progression was due to the Creator who worked "in the manner of natural law" was a meaningless statement. Huxley was trying to separate a scientific proposition (whether or not the fossil record was progressive) from metaphysical speculations (i.e., the Creator worked in a lawlike fashion). But in the *Vestiges*, progression was inextricably linked with the presence of a Creator and theological arguments were mixed with scientific ones. This ensured Huxley's wrath. In addition, the book was filled with "blunders and mis-statements" giving a totally distorted view of the fossil record.[25] Thus, Huxley's ire was not directed at the idea of transmutation, but rather at using scientific facts (which were often wrong) to promulgate a religious belief.

According to Huxley, what appeared to be a scientific discussion on the organic world was nothing more than an elaborate version of the Book of Genesis. However, Huxley's vitriolic attack of the *Vestiges* is doubly ironic. First his antitheological rhetoric leaves him open to the same criticism he has of his opponents. And second, while most scientists claimed it was the poor science of the *Vestiges* that they objected to, underneath their professional criticism was the fear that the naturalistic explanation of creation threatened the special status of humanity and their most deeply held religious views.[26]

In struggling to free science from theological implications, Huxley let his own philosophical beliefs influence his interpretation of the data. However, Huxley was not unique in this respect.[27] He was also guilty of distorting his opponents' views. For example, Owen made no reference what so ever to the argument from design in his critique of Lyell's antiprogression stance in his Presidential Address. But claiming that men such as Owen, Buckland and others were letting their theological beliefs influence their science served as a powerful rhetorical strategy for discrediting their views. Darwin's theory had tremendous appeal to Huxley in spite of doubts over significant aspects of it because it struck a serious blow to the argument from design.

Evolution, like the other case studies in this book, was subject to the same kinds of issues that influence the production of scientific knowledge. Even if scientists are biased, misinterpreting results because of certain theoretical preconceptions, or because of political or other types of agendas that they are pushing, ultimately nature asserts herself. Evolution became a powerful theory for the life sciences, not because it turned out to be correct (indeed as evolution continues to generate new ideas, many of them are actively disputed and some border on pseudoscience), but because it was a hypothesis that could be tested. Huxley did not think that the transmutation hypothesis had been proven, but he recognized that it was a powerful research tool. Therefore, he championed the cause of evolution in spite of his disagreements. Even when Huxley thought that the facts of nature either contradicted the theory (as in his early views concerning gradualism and the fossil record) or that at the time not enough evidence existed to move the theory from the realm of hypothesis to the realm of fact (as he thought was the case with natural selection), these areas were open to investigation. They potentially could be proved or disproved. Furthermore, the mid-nineteenth century was a particularly exciting time for biology. The cell theory of Schleiden and Schwann, embryological research, improvements in microscopy, the truly dramatic finds in paleontology, and an enormous increase in data on the distribution of plants and animals, all provided evidence for Darwin's theory that earlier theories about the history of life lacked. The truly monumental discoveries in virtually every area of biology, but especially paleontology meant that the fact of evolution could not be denied. Fossils demonstrated the reality of evolution, even if the details of how the actual change occurred remained controversial.

Huxley often chose to emphasize "the facts" in his defense of evolution, but he also claimed that there was "no field of biological inquiry in which the influence of *The Origin of Species* is not traceable." He immediately recognized the power of the Darwinian hypothesis predicting that "as the embodiment of an hypothesis it is destined to be the guide of biological and psychological speculation for the next three or four generations."[28] Huxley's words were prophetic. One of the hallmarks of a good scientific theory is that it should continue to generate hypotheses that can be tested. Unlike the other case studies in this book, evolutionary theory has proved to be exceeding robust in this regard. The very issues that most interested Huxley have been a rich source of new research as Darwin's basic ideas continue to be debated and extended.

The issue of saltation has continued to be revisited throughout the twentieth and twenty-first century, receiving its most serious hearing with the theory of punctuated equilibrium promulgated by Niles Eldredge and Stephen Gould.[29] Today one of the most exciting areas of research is evolutionary developmental biology or "evo-devo" that is attempting to fully integrate the findings from development with the research in evolution. Recent research in developmental genetics has wreaked havoc with the traditional distinction between homology and analogy because "homologous" genes are responsible for "analogous" processes. Not only is development the key to understanding how a single cell becomes a complex, multibillion celled organism, but also it is thoroughly connected to evolution since it is through changes in the embryo that changes in form arise.[30] While it is true in the most general sense the fossil record is progressive, the debate over progression has been reframed today as evolutionists examine more carefully what they mean by complexity. It would be quite difficult to prove that a modern-day lizard is more complex than a dinosaur. How can complexity be measured? Is diversity a measurement of complexity? Has diversity increased or decreased over time or is it cyclic?[31] These are just a few of the questions that are being examined by modern evolutionary biologists.

These disagreements and disputes among evolutionists are the marks of a good scientific theory. *Thatige Skepsis* should be the motto of every scientist. Evolution has proven to be an extremely powerful theory, but like all science, it is not immune from the same kinds of problems that have plagued the marginal sciences. The final chapter explores how an area of investigation merits the word "scientific" and examines some modern-day disputes that connect to the case studies, in which not only the interpretation of observations are contested, but their status as legitimate science as well.

7

Negotiating the Boundaries of Science

An Ongoing Process

The credible is by definition, what is believed already and there is no adventure of the mind there.

—Northrup Frye, *Words with Power*

The Culture of Science

SEA SERPENT INVESTIGATIONS, phrenology, and spiritualism never became part of the scientific mainstream. However, the boundary disputes surrounding these marginal sciences of the Victorian era have very much shaped the practice of science today. Although boundary disputes remain, they are not the same in character at least within scientific disciplines. Scientists seemed to have learned something of a general nature about what science can and cannot do. For example, while there is still speculation about mind and body, the scientific community is no longer investigating after-death communication with spirits or trying to find proof for the existence of a soul. The failures of phrenology/physiognomy have meant that any claims based on external anatomy to explain mental phenomena are regarded as highly suspect. Although a strong amateur tradition remains in the gathering of fossils, the Victorian period resulted in the full professionalization of paleontology and geology where training and proper credentials are essential to having one's work regarded seriously. Darwin's theory of evolution emerged as the great unifying theory of biology. In the words of geneticist Theodosius Dobzhanksy, "nothing in biology makes sense except in light of evolution." Nevertheless, evolutionary theory continued to be enmeshed in controversy throughout its history as the science of evolution itself evolved.

In Darwin's time, descent with modification was relatively quickly accepted, but natural selection was problematic. No well-grounded theory of heredity existed to explain the source of variation that provided the raw material for natural selection to act on. Yet Mendelian genetics was initially used to argue against natural selection and it was not until the rise of population genetics that natural selection was finally vindicated. However, other issues remained. The question of gradual versus saltational evolution along with the relationship of microevolution to macroevolution, and the integration of developmental biology with evolutionary theory have led to some very contentious debates among evolutionary biologists. Human evolution continues to be a highly controversial topic both within the scientific community and for the general public. Finally, the Argument From Design, which was so prominent in nineteenth-century debates over evolution and the history of life, has reappeared with a vengeance in the United States. Intelligent Design has made serious inroads into the general public's perception of the status of evolution as a scientific theory, which in turn is influencing the discussion in the scientific community. The disagreements surrounding evolution are many, and are not just over the quality of the evidence, or whether a particular idea is right or wrong, but whether certain ideas even deserve to be considered "scientific."

The case studies, thus, each in their own way dramatically demonstrate that it is very difficult in the short term to determine what scientific ideas are going to prevail in the long term. At the time each of these topics attracted the attention of eminent scientists, illustrating that the boundary between science, marginal science, pseudoscience, and myth is not sharp. How is it decided that an idea or area of investigation merits the word "scientific" attached to it?

Philosophers of science have long been interested in trying to characterize science in such a way that differentiates it from other kinds of activities. None of their attempts have been entirely satisfactory. Inductivism, positivism, hypothetical-deductivism and falsificationism along with a variety of other proposals failed to achieve demarcation between science and nonscience based on a priori principles. Yet in a very practical way demarcation between science and nonscience is accomplished every day—from the designing of science curriculums to deciding what research receives funding. Ptolomy is taught in a history of science course, but is no longer considered a viable theory to be debated in an astronomy class. The National Science Foundation funds research in chemistry, but not alchemy, on the existence of extraterrestrial life, but not on the existence of Bigfoot. Research concerning the strange world of the quantum is deemed legitimate, but investigations into the strange world of psychic phenomena are not. Are there some basic ground rules that frame the negotiations that might help explain why modern science has been such an extraordinarily successful enterprise? In spite of the limitations of the various philosophical proposals, most scientists believe that the essence of science lies in its method, which consists of five basic steps:

1. Observe and collect data.
2. Induce general hypotheses or possible explanations from the observations.
3. Deduce specific corollaries that must also be true if the hypothesis is true.
4. Test the hypothesis by checking out the deduced implications.
5. Do repeated tests or develop new ones.

Nevertheless, this scheme is problematic. Although the scientific method acknowledges the importance of speculation, it suggests that if a hypothesis is not borne out by experiments then the theory is thrown out. But, scientists more often than not hold on to hypotheses in spite of experiments failing to confirm them. Virtually all good theories, especially in their early stages, are underdetermined by anything that might be called evidence; "facts" are underdetermined as well. Francis Crick provocatively remarked that you should build your theory on as few facts as possible. Why? Because in the early stages of an investigation, you don't know what is a correct fact and what is an incorrect fact. The scientific method assumes that a clear distinction exists between theory and observation; however, what counts as theory and what counts as data is continually being revised. For example, Darwin's theory made sense of an enormous number of facts, but it was also quite speculative. He claimed that species change over time by the accumulation of tiny variations. However, such gradual change was not observed in the wild and the fossil record certainly did not support such a view. Rather, it was full of gaps with species abruptly appearing and disappearing. Moreover, the reconstruction of extinct organisms was often based on very few fossil bones. Fossils also point out another limitation of defining science in terms of the experimental method. It excludes sciences such as paleontology and astronomy that are observational and historical. Although scientists generate hypotheses about the history of the universe and life on earth and look for evidence that corroborate those hypotheses, they cannot do experiments on the past—it is gone.

The most important limitation of the scientific method as a description of science is that it leaves out the social aspect of doing science. Scientists are humans, with biases, ambitions, and prejudices. This is not to be interpreted as a criticism of science, nor is it a claim that evidence is unimportant in how scientists choose between alternative interpretations of the world. Simply put, the scientific method is a bit of a myth. It is an ideal that scientists ascribe to, but it does not reflect how scientists actually work, particularly in areas that are at the frontiers of scientific research. The case studies in their historical context exemplify frontier science.

A more accurate description of how scientific knowledge accumulates is Henry Bauer's notion of a filter.[1] Human knowledge comes from all kinds of people, based on their observations and suggestions. It is a mixture of hunches, superstition, folklore, observation, conventional wisdom, wild ideas, and is influenced by people's prejudices, ideology, competence, persistence and all the other traits both good and bad that make us human. In the early stages of frontier science a wide range of ideas are considered and forms the basis for scientific speculation. The existence of sea serpents became credible in light of recent fossil discoveries. The discovery and elucidation of the laws governing the quite mysterious forces of electricity and magnetism lent plausibility to the existence of an undiscovered psychic force and suggested that it might finally be possible to investigate the existence of the spirit scientifically. Other advances in natural knowledge led Gall and his followers to claim that it was now possible to truly have a science of the mind. The fossil record, the distribution of plants and animals, embryology, and the artificial selection of plants and animals suggested to Darwin that species were not fixed.

At this stage of the investigation knowledge is highly unreliable, but various filtering processes winnow the possibilities of what will eventually become accepted as scientific knowledge. First, new ideas must be tested against what are accepted ideas. Although ideas about earth history were in flux, a general consensus was emerging that the history of life was progressive. The existence of the sea serpent as a relic from the past went against that consensus. Spiritualism was the province of fraudulent mediums, outlandish claims, unrepeatable results, and most scientists did not think it was possible to scientifically investigate the nonmaterial world. Phrenology had the opposite problem in its early stages. It was attacked as materialist, doing away with the mind/body dualism of Descartes. In doing so it challenged the authority of philosophers who maintained that introspection was the path to reliable knowledge about the nature of mind. Darwin's theory also was regarded as materialistic. It contradicted the long held view that species were fixed, created by God, but that idea had already been challenged on several different fronts.

In the next stage of the filtering process, a variety of factors contribute to the following a particular scientific idea may generate. Even advocates for the strict objectivity of science admit that explaining why theories enjoy the popularity they do at a particular moment in history is extremely difficult. Scientific respectability is not determined just by the nature of the "evidence" because how evidence is interpreted is a complex and multifaceted affair. Ideological issues are often inextricably intertwined with scientific ones. Prior theoretical and/or psychological commitments, the politics of professionalization, and cultural norms about morality and the human condition all play a role in determining scientific credibility.

Theories that reinforce the prevailing societal values will more readily find acceptance. Spiritualism had tremendous appeal to a public who was enamored with scientific advances, but was anxious about the materialistic implications of this new knowledge. Scientific proof of the spirit world would provide a solution to these conflicting emotions. Phrenology may have been attacked initially as materialistic, but physicians quickly recognized the potential of phrenology as a means to increase their prestige and status in society. For the general public it offered a scientific prescription for self-help and improvement.

The case studies also highlight another crucial aspect of the filtering process, which is the actual creation of boundaries. Scientists are able to advance their professional goals by creating boundaries between their activities and those they define as nonscience. This includes gaining cultural authority and career opportunities, and obtaining funding for their projects.[2] Thus, in the emerging disciplines of paleontology and geology, making a distinction between amateur and professional was one way of creating a boundary. Most people that were interested in the sea serpent were amateurs. In addition, the sea serpent was too closely allied with myth and legend. Although Lyell hoped the sea serpent would provide support for his steady-state theory regarding earth history, he like Owen also was committed to the professionalization of geology and wanted it recognized as a true science. With such an agenda, sea serpent investigations would remain marginalized.

Philosophers by labeling phrenology as pseudoscience hoped to maintain their monopoly on theories of mind. In addition, most antiphrenologists were High Church Anglicans and followers of Paley's natural theology. They had more personal reasons for wanting to keep phrenology out of the universities, a doctrine that they also regarded as materialistic and atheistic. Darwin and his followers also recognized that creating distinct boundaries between science and theology was critical to their advancing their ideas not only in the university, but also society at large. Huxley's defense of evolutionary theory served as a vehicle to attack theology and to further his broad campaign to transfer the prestige, power, and moral authority enjoyed by the church to the temple of science. It was not an accident that his Working Men's Lectures were called Lay Sermons. Michael Faraday and John Tyndall echoed Huxley's views, both men giving many popular lectures. Their lectures, like Huxley's, were intended to do more than just inform the public about the latest scientific advances. Rather, public advocacy of "professional science" and the defense of evolution went hand and hand, and were used to marginalize natural theology as evolutionary theory eventually became the new orthodoxy. As the scientific naturalists gained control of the university establishment, this would also mean there would be no place for the investigation of spiritualism within the scientific mainstream.

Public advocacy and professionalization of science also had one other vitally important function—economic. Very few people were earning their living as scientists. Huxley, in spite of being in many ways extraordinarily successful, always struggled financially. In 1873 he complained, "Science in England does everything—but pay. You may earn praise but not pudding."[3] Huxley wanted more resources devoted to the scientific enterprise. This included more employment opportunities, a strengthening of the science curriculum in schools and universities, and increased government and private patronage for research facilities. He was a tireless advocate for not only having the natural sciences being taught in the schools, but that the courses should involve actual laboratory practice to train future scientists. Physiology deserved the status of a true experimental science, and should not only be confined to the medical school curriculum. Professional scientists, who were entitled to be paid for their knowledge and their time, should do both the research and teaching of physiology.

Huxley's advocacy of physiology to be part of university curriculum highlights what is the final component in defining scientific knowledge. In the short term a virtual cornucopia of claims, observations, experiments, and interests are contributing to whether something is regarded as scientific. However, the final and critical phase of the filtering process that eventually results in dependable scientific information in the long term is that science is fundamentally a collaborative endeavor. Physiology had accumulated a body of knowledge over hundreds of years based on repeated experiments and observations, with past investigators often disagreeing. Scientists rarely refute their own hypotheses, but that doesn't matter because theories are commented on by many other scientists working in the field. Results must be replicated. New findings must not contradict what was previously regarded as "fact," and ideally also extend the explanatory power of a particular theory. In other words, facts and theories must endure the scrutiny of members of the scientific community for an extended period.

In the case of the sea serpent, ultimately the kinds of objections that Richard Owen raised were definitive. It doesn't matter that he clearly used the sea serpent controversy as a vehicle to push his institutional and political agenda. In spite of all the searching, no physical evidence of a serpent turned up. As the century proceeded, more and more fossils were found, but nothing in the more recent strata was anything like a plesiosaur. The lack of evidence indicated that these great "serpents" of the past had indeed gone extinct and brought into serious question the present-day eyewitness accounts. Lyell's belief in the existence of the serpent was tied to his theory of climate, but he eventually realized that the existence of the sea serpent as a relic from the past was actually counter to his theory. The weight of the evidence simply argued against the existence of sea serpents and the scientific community did not think it was worth spending either their time or their limited financial resources in pursuit of these highly elusive creatures.

It can never absolutely be proven something does not exist, whether it is sea serpents, spirits, or flying saucers. As someone remarked to me, sea serpents were the UFOs of the nineteenth century. A judgment must be made, which is what the scientific community did. This did not mean that every serpent sighting was accounted for, but it did mean that all things considered, including the lack of specimens and what was known about the history of life from the fossil record, it was extremely unlikely that a relic from the Cretaceous survived to the present. The same was true of spiritualism. Some phenomena at séances remained inexplicable. However, the vast majority of phenomena could be explained without resort to spirits or an undiscovered force. Furthermore, virtually all the mediums were exposed as charlatans. The scientific community passed judgment. Not only were most of the phenomena bogus, but also the immaterial world could not be investigated in a meaningful way. Scientists chose to apply their knowledge, skills, and resources to other questions. Nevertheless, the existence of a spirit world continues to periodically attract the attention of iconoclasts that have impressive scientific credentials.

At the turn of the century, physics underwent a profound revolution with the development of relativity theory and quantum mechanics. Atoms were no longer indivisible unalterable building blocks of matter. The ideas of Bohr, Einstein, and Heisenberg brought into question basic assumptions about causality, materialism, and determinism, thus providing a possible framework in which psi phenomena might make sense and gave a new burst and sense of legitimacy to investigations of the paranormal. Nevertheless, after decades of disappointing results with numerous exposures of hoaxes and fraud, parapsychology still remains at the margins of scientific research.

Phrenology had a more complicated trajectory, but also illustrates the importance of science as a cooperative endeavor in the interpretation of evidence. Its basic ideas were incorporated into mainstream research in psychology and neurobiology, but most of its detailed claims turned out to be false. This in itself was not the reason for phrenology's marginalization since many scientific ideas eventually are shown to be wrong. In spite of the scientific naturalists gaining control of the universities, phrenology with its biological approach to the study of mind did not become part of the curriculum for good reasons. Although debate and controversy characterized the early stage of phrenology, this did not continue *among* the phrenologists themselves. Rather, phrenology lost essentially all of its reputable medical and scientific support because contradictory evidence was ignored. Phrenologists did not do the kinds of careful experiments that were needed to build on the basic ideas of Gall. Those aspects were absorbed into the research agenda of the developing field of neurobiology. Instead, phrenology became essentially a social movement. Rather than marginal science, a more apt description is nonscience. However, nonscience that claims to have the authority and power of science when it is not warranted should be considered pseudoscience. Combe did not base his reform movement on moral principles, but rather on the supposedly "objective," "scientific" principles of phrenology. It was not just that the evidence for phrenology was poor, but that a reform movement was being called science.

Social policy is not the same thing as science, even if it is based on sound scientific principles. Thus, the decision to universally inoculate children against polio is as distinct from the science responsible for developing the vaccination as the decision to send a man to the moon is distinct from the science that made such a project feasible. Scientific findings can and should be used to guide policy decisions, but particularly when it comes to social policy, moral and ethical concerns are also important and may be in conflict with scientific findings. Science influences and is influenced by the culture in which it is embedded.

The problem of distinguishing between claims that have scientific validity and claims that someone says have scientific validity is not unique to the marginal sciences. This is at the core of defining what is considered good science and is continually being reevaluated. Although eventually the evidence simply did not support the existence of serpents and spirits or the claim that a bump exists on the skull that corresponds to the trait of acquisitiveness, this was not apparent in the early stages of these investigations. More importantly, an array of social, political, and moral factors shaped the discourse surrounding these controversies.

This continues to be true today. Science does not occur in a vacuum as particular ideas become accepted or rejected due to a complex array of issues. In the remainder of the chapter I examine two cases of science in progress that connect to the historical case studies. The first one explores the role that fossils continue to play as possible links between myth and science as humans remain fascinated with the monsters of mythology, hoping to find concrete examples of their existence. The second examines evolutionary psychology, and just like phrenology claims to provide a definitive scientific explanation of human nature. Both of these investigations are mired in controversy and illustrate the boundaries of science are continually being negotiated.

Of Plesiosaurs and Protoceratops: The Meaning of Fossils Revisited

The discovery of fossil "monsters" in the early part of the nineteenth century was the crucial piece of evidence that suggested that the sea serpent might not just be a creature of ancient legend. As W. J. T. Mitchell has written, "Real fossil bones are inevitable occasions for imaginative projection and speculation."[4] That speculation is shaped by scientific practices concerning the collection of bones, the organization of digs and expeditions, and the cleaning and reconstruction of bones that finally results in a description and classification of fossil objects. However, the story that is told about particular fossils at a particular moment in history is "tied as well to representations of the world and changing social and cultural configurations."[5] Today fossils could potentially provide a bridge between mythology and science in the new role they are playing in the interpretation of various legends. In doing so the boundary between paleontology, archeology, and anthropology is blurred, as "evidence" is reinterpreted.

In *The First Fossil Hunters*, Adrienne Mayor suggests that the discovery of huge fossilized bones were the origin of many Greek and Roman myths.[6] Fossils as connecting links between archeology and natural history was implied much earlier in the writings of Robert Hooke (1635–1702). Just as Roman coins or urns are clues to interpret a past civilization, fossils are the "medals, urns or monuments of nature." This is a particularly fitting analogy for Mayor's work. Mayor is not a paleontologist, but a folklorist and she has suggested that fossils provide clues to understanding and interpreting civilization. However, she has turned the analogy on its head by using the urns and monuments of civilization to support her thesis. As she writes, "The tasks of paleontologists and classical historians and archeologists are remarkably similar—to excavate, decipher, and bring to life the tantalizing remnants of a time we will never see."[7] Using evidence collected from personal interviews, trips to Greek and Italian museums, and studies of classical myths, she argues that the discovery of large fossils were the basis for many of the monsters of Greek and Roman mythology.

Based on evidence from ancient historical and literary sources, Mayor thinks that what has been labeled as descriptions of mythical creatures today can be identified as fossil remains of various animals. The Monster of Troy vase made in Corinth in the sixth century B.C. depicts Heracles and Hesione confronting the monster of the Homeric legend that appeared on the coast of Troy, near Sigeum after a flood. Classical art historians have

FIGURE 7.1. Griffin and Baby. Drawing by Adrienne Mayor. Originally published in Adrienne Mayor, *The First Fossil Hunters: Paleontology in Greek and Roman Times* (Princeton, NJ: Princeton University Press, 2000). Reprinted by permission of Adrienne Mayor.

always assumed it was a sea monster (although in Homer it is described as a land creature), but it does not look like a typical sea monster as represented in other artifacts. In addition, while the other animals and humans on the vase are well rendered, the monster is crudely drawn. Upon more careful examination Mayor has suggested that rather than a sea monster coming out of a sea cave, instead it appears to be a head protruding out of a cliff. It is suggestive of a large skull eroding out of an outcropping. The area around Sigeum is rich in fossil deposits and several paleontologists agree that the shape and size of the skull was consistent with that of a large Tertiary mammal emerging from an outcropping.[8] Mayor discusses several other examples of fossils represented in paintings and on pottery from all over the ancient world that depicted various myths. However, the question remains were these fossils the basis for the legend or were they interpreted in the context of a particular legend that already existed?

An example that has captured the attention of many paleontologists is the griffin (see figure 7.1). Since the seventeenth century the scholarly world from archeologists to zoologists has thought the griffin was an imaginary construct, a combination of a lion and an eagle, but Mayor claims it was no simple composite such as Pegasus or the Sphinx. Although many other monsters dwelled in the mythical past, the griffin was believed to exist in the present day. Images of griffins were found with ancient Greek heroes and gods, but they were not the offspring of gods, and were not associated with the adventures of Greek gods or heroes. Griffins had no supernatural powers. Supposedly ordinary people encountered them while prospecting for gold in Asia. For Mayor the griffin seemed a prime candidate for a paleontological legend.[9] The griffin was found on Greek and Roman vases, and one was tattooed on the skin of a fifth century B.C. mummified nomad found on the slopes of the Altai mountains on the edge of present-day Mongolia. In 360 B.C., the physician Ctesias wrote, "gold originated from the high mountains in an area inhabited by griffins, a race of four-footed birds, almost as large as wolves and with legs and claws like lions."[10]

Mayor has suggested that the griffin is in actuality the dinosaur Protoceratops (see figure 7.2). A relatively small dinosaur, only about six feet long, it had a great massive beak dominating its head, making it look quite

FIGURE 7.2. *Protoceratops*. Photograph by R. C. Andrews. Reproduced by permission from the American Museum of Natural History.

eaglelike. The head passed back to a bony frill that might have been described as the griffin's wings and the rest of its body looked like many quadrupeds with a long tail. In silhouette, the griffin and the dinosaur look remarkably alike. Fossils of Protoceratops are quite abundant and a 1920 expedition to the Altai Mountains discovered over one hundred individuals. The ancients undoubtedly knew these fossils since they were so common, and they have been found in the same areas of gold bearing deposits. They were indeed guardians of great wealth. Whether the tribesmen deliberately encouraged the legend of the griffin to protect these sites is quite speculative. Paleontologist Paul Sereno disputes the whole idea, claiming that the fossils are found several hundred miles from the gold deposits.[11] However, paleontologist Mark Norell disputes this, claiming that fossils have been found relatively close to gold deposits in the Gobi Desert. Nevertheless, he also points out that it is problematic to assume that the griffin of classical antiquity and the griffin tattooed on Bronze Age nomads represent the same creature. It is like comparing Asian and European dragons. Representations of griffins exist in all kinds of shapes, sizes, and proportions. "Some look like Protoceratops—others look like dinosaurs that were not found (at least in the sense of by paleontologists) until the late 1900s in North America. Certainly these would not have been known to people in the Mediterranean or Asia."[12] Nevertheless, the theory of griffin as dinosaur seems plausible.

In 1868 Thomas Huxley suggested that dinosaurs were the common ancestors of birds and reptiles. Today, with more evidence available than in Huxley's time, dinosaurs and birds are classified hierarchically. Both are a kind of reptile, with birds evolving from a particular lineage of dinosaurs, most likely from theropods. Griffin as dinosaur represents the convergence of myth with reality. Although many paleontologists have found Mayor's thesis compelling, Sereno is quite hostile to her ideas referring to them as "sophomoric" and her interpretation of the griffin as "pure fiction." He maintains that no people in the history of the world have ever come up with an interpretation of something as unusual as a frilled dinosaur from fossils in the ground, arguing that what looks obvious to us now would not be at all obvious to the ancients. Mayor, however, claims that scholars throughout history have consistently ignored or misinterpreted a wealth of different kinds of evidence, from written texts to pottery to paintings to fossils that make a strong case in support of her thesis. Norell agrees that plenty of evidence exists from all over the world of fabulous stories based on the misinterpretation of fossils and does not even think the griffin case is the most compelling. However, Sereno accepts the received view of the scholarly community who has not made this link between fossils and myth. More importantly, he bases his opinion in part on his experience with some of the last surviving nomads, the Tuaregs of

Africa. Although an abundance of dinosaur bones exist in the nearby desert, no "monsters" are found on any of the cliff drawings and engravings of these people. One Tuareg nomad interpreted the bones as belonging to "large camels." For Sereno this was a perfectly reasonable explanation as that was the only large animal they had ever seen.

Sereno is committed to scientific methodology and a hypothesis should be capable of being tested and make predictions. If fossils gave rise to griffins, then a culture that has no fossiliferous outcrops should have no griffins. But one finds representations of griffinlike beasts all over the world, particularly in South America. For Sereno, the more reasonable explanation is that they resemble combinations of revered animals of that day such as a lion and an eagle. Conversely, cultures found in areas such as the Sahara that were rich in fossils should have given rise to myths of dinosaur-like creatures, but they did not. Instead, Sereno argues that traceable ideas about fossils as unusual as dinosaurs lead not to griffins, but to huge humans or overgrown mammals. However, as paleontologist Norman McLeod pointed out, we do not know nearly as much about native South American mythology as that of the ancient Greeks and Romans. There are many frilled fossils in South America, and he thinks it is quite possible that a legend was associated with them. It might have been lost for a variety of cultural factors, including few written records.[13] This would be even truer for the nomads of the Sahara.

Sereno's most basic objection is that it is very difficult to recognize fossils and that Mayor is reading our modern knowledge back into the past. He claims that most fossils like dinosaurs are not "easy." "They are complex to the eye, impossible to extract from the ground, difficult to determine what is bone and what is not."[14] Inexperienced paleontologists will often destroy the very fragile fossils before they remove something from the ground like a Protoceratops skull. Jack Horner, however, argues that this is not always the case. White skeletons weathering out of red sandstone would be very difficult to miss. On his last expedition to the region with a crew of five, he found a nearly complete Protoceratops skeleton on an average of one every hour or so. He thinks Mayor will turn out to be correct claiming that "[t]here is very little about the morphology of the griffin that cannot be imagined from the Protoceratops."[15] Norell also disagrees with Sereno. Not only has he found (and mentions that Sereno has also had similar experiences) fossil dinosaur skulls sticking right out of the ground, he also disputes that nomads from ancient times would not know what they were seeing. "Nomadic people and people in general up to about two hundred years ago had a much keener awareness of animal bones than we do now. They hunted and butchered their own meat. Many nomads and traditional people that I have encountered in different parts of the world can identify all sorts of animals from just a few bone fragments. They certainly would be able to tell if a skeleton was one that they were familiar with, saw every day or was something completely different."[16]

Whether Mayor's ideas will prevail or not is an ongoing topic of discussion among paleontologists, anthropologists, archeologists, and classicists. In spite of Norell being the cocurator of a major exhibit on mythical creatures at the American Museum of Natural History (discussed later) that prominently features the griffin and suggests a possible link with the fossil Protoceratops, when pushed as to whether he thinks Protoceratops is the origin of the griffin myth his reply was, "Although it is an interesting idea, I do not think so."[17] Once again the meaning and interpretation of fossils is enmeshed in controversy. However, such controversy is inevitable. Just as in Victorian times, a variety of factors are shaping the debate, some of which are similar and others that are historically contingent. Throughout this book, we have seen how fossils were used to support vastly different interpretations of the history of life, which in turn reflected the social, religious as well as scientific issues of the day.

In the nineteenth century a small, but influential group of British paleontologists argued that the fossil record provided evidence of the flood as told in the Bible. At the same time findings from geology unequivocally demonstrated that the earth had to be far older than what the Bible suggested. Fossils also documented the reality of extinction, Nevertheless, the vast majority of geologists and paleontologists believed the progressive nature of the fossil record provided evidence in support of a Divine Plan, which reflected a deep-seated desire to use the authority of science to reinforce the authority of religion.

Today, the Argument From Design has reemerged as Intelligent Design, and is framing the public debate over evolutionary theory. This development provides us with a clue to Sereno's hostility to Mayor's ideas. Sereno is deeply committed to science education. With Gabrielle Lyon, he cofounded Project Exploration, an organization dedicated to bringing dinosaur discoveries and natural science to the public and providing innovative educational opportunities for city youths. As Intelligent Design continues to make inroads and threatens to seriously compromise the quality of science education, the need to make the distinction between science and other kinds of endeavors is felt acutely by those people who are truly in the trenches in this fight. Academics may speculate about the links between myth, history, and science, but these ideas are easily and often distorted. Finding fossils, excavating them, and reconstructing them is hard work. Today, all kinds of tools are used in a restoration, from studying the laws of biomechanics to taphonomy.[18] Sereno wants the public to recognize the hard science that goes into interpreting fossils and also to realize that science is not the "truth," but is about hypotheses that undergo continual refinement and tests. In this way science is quite distinct from myth or legend as well as religion.

Paleontology has come a long way from its origins as primarily a historical and observational science. Establishing phylogeny (i.e., whom is descended from whom) is now based on cladistics, a sophisticated mathematical analysis of the distribution of many different characters. In dinosaurs, for example, characters such as the shape and placement of the bones that make up the pelvis, the shape of the tail, of the head, the bone structure of the forelimb, and of the feet are just a few of the traits that are examined. When a new dinosaur is found it is added to the analysis and may corroborate the current hypothesis, but it also may not. As more and more fossils are found a hypothesis may be generated that better fits the evidence. Thus for example, while most paleontologists think birds evolved from theropods, this was not always so. Until relatively recently the prevailing view was that thecodonts gave rise to birds. With the discovery of the feathered dinosaurs in the 1990s, the evolution of birds remains an active area of research and has also resulted in a major rethinking about dinosaur evolution.[19] Furthermore, paleontology is about more than just constructing phylogenies. What other kinds of evidence might be relevant to understanding the past and the creatures who inhabited it? Although many mythological creatures are purely the product of the imagination and have nothing to do with paleontology, can the study of fossil legends contribute to our understanding of fossils? Or do "legends of fabulous beasts represented by fossil bones have little ethnological and no paleontological value," as George Gaylord Simpson wrote in his dismissal of Native Americans' contribution to paleontological history.[20]

The folklore of paleontology is a new field of study and as such is vying for respectability. As an emerging discipline it will inevitably be caught up in boundary disputes. Sereno, like Simpson is vested in having paleontology being recognized as a rigorous science, its boundaries distinct from those disciplines involved in cultural studies. However, a culture's myths concerning the history of the earth provides a context for studying traditional accounts that do explicitly refer to fossil discoveries. Mayor believes that examining fossil legends offers a new way of thinking about pre-Darwinian encounters with prehistoric remains. Unfortunately, as paleontologist Tim Tokaryk points out "the only way to empirically demonstrate a union between fossils and mythology, is [to examine] the cultural handling of the fossils themselves, and this [information] is admittedly sparse."[21]

As paleontologists subject dinosaurs to increasingly sophisticated scientific analysis, they simultaneously play a significant role in the creation of the myths and imagery that surround the object of their study. Sereno may believe that "human minds have always been far more imaginative [than having to] rely on scrappy fossils for their images"; however, the dinosaur had a mythical quality right from its beginning. Richard Owen first coined the term, which means terrible lizard, to describe the group of large extinct reptiles. The interpretation of dinosaur fossils has and continues to further the disciplinary agendas of professional geologists and paleontologists. Inventing the dinosaurs and putting them in their own taxonomic category insured that Owen's name would be forever associated with these huge monsters rather than people such as Gideon Mantell. Mantell's reputation soared with his discovery of the iguanodon and his creation of "The Age of Reptiles" with his highly successful book *The Wonders of Geology* (1838). More importantly, the dinosaur became a new icon of the museum display, which has lasted to the present day. Richard Owen collaborated with artist Benjamin Waterhouse Hawkins who created life-size and lifelike sculptures of various dinosaurs at Sydenham Park for the Crystal Palace exhibit of 1854. On New Year's Eve 1853, Hawkins hosted a dinner party for Owen and other dignitaries in the belly of his half completed model of the iguanodon (see figure 7.3). It provided an opportunity for Owen and Hawkins to strengthen their scientific authority of their work. On the basis of a few bones and some teeth Owen decided that the iguanodon was more mammalian in both its anatomy and physiology. Rather than Mantell's reconstruction of a giant creeping lizard, Owen transformed the iguanodon into a huge rhinoceros-like form standing on four massive legs. Owen, in spite of later disagreeing with the antiprogressive stance of Lyell, wanted to refute the growing interest in Larmarckian evolution that he believed was threatening to science and society. Suggesting that this ancient creature was more similar to present-day mammals rather than reptiles was counter to the progressiveness intrinsic to Lamarck's ideas regarding the history of life.[22] The park was a sensation. Owen wrote a guidebook for it. Forty thousand people including Queen Victoria eventually came to see the dinosaurs, wading in man-made lagoons, climbing an antediluvian tree, surrounded by mock geological strata.[23] In the Victorian age museums emerged as places not merely to entertain, but rather to educate the public. Yet who could deny the entertainment value of the dioramas, panoramas, and 'peep shows' along with the displays of the huge fossils as museums created their own narratives of the antediluvian age. "They presented the prehistoric world as exhibition."[24] Engaging the public with this sense of wonder, the dinosaur helped build support for Owen's museum program and also encouraged further overseas explorations.[25]

Dinosaurs are neither terrible[26] nor lizards and the term refers to an artificial and scientifically incoherent taxonomic category. Yet dinosaurs remain an essential component of any natural history museum and the larger ones employ paleontologists along with many other professional scientists. Paleontology is no longer struggling for legitimacy, but it ranks relatively low in the hierarchy of science as it competes with other sciences such as

FIGURE 7.3. Dinner in the Iguanadon. Dinner held in the Iguanodon Model at the Crystal Place Sydenham. From the *Illustrated London News*, 1854. Published in M. J. S. Rudwick, *Scenes from Deep Time: Early Pictorial Representations of the Prehistoric World* (Chicago: University of Chicago Press, 1992). Reprinted by permission of the publisher.

molecular biology and genomics for funding. Since most funding comes from the government, a way to increase funding for paleontology is to engage the public in its work.

The public has a fascination with dinosaurs, but that fascination, just as in the Victorian era, has been fueled in part by the professional community in an effort to sell paleontology. Many of the dinosaur exhibits such as Sue at the Field Museum in Chicago are outstanding and have done an excellent job in trying to educate the public about evolution. But many other dinosaur exhibits have a Disneyland-like quality to them. The myth making continues in order to help secure funding for the nuts and bolts research not only in paleontology, but other disciplines as well. Norell admits, "Without the mass appeal of dinosaurs, my institution and others like it certainly would not be in the position to support world class genomics, conservation and molecular programs."[27] Sereno and Horner have both contributed to the mythology. Horner was the paleontological consultant for the movie Jurassic Park and the paleontologist Dr. Grant was loosely modeled on him. A visit to Sereno's website portrays Sereno as the intrepid dinosaur hunter, overcoming immense difficulties, working in stifling conditions. He has been called the "Indiana Jones of paleontology," and he admitted that he had become "the virile dino-boy of the moment."[28]

Linking fossils to mythology is another way to sell paleontology to the public. A 2006 exhibit of Chinese dinosaurs at the Field Museum featured a complete Protoceratops fossil and mentioned that Protoceratops might be the origin of the griffin legend. This is a highly controversial suggestion, yet it was chosen to be the key piece of information conveyed in the brief caption of the fossil. The American Museum of Natural History in a major exhibition in 2007 entitled "Mythic Creatures, Dragons, Unicorns and Mermaids" also linked griffins to Protoceratops. By the display of the fossil Protoceratops was the caption:

> We stopped at a low saddle between the hills. Before I could remove the keys from the ignition, Mark sang out excitedly. . . . Several feet away, near the very apex of the saddle, was a stunning skull and partial skeleton of a Proto[ceratops], a big fellow whose beak and crooked fingers pointed west to our small outcrop, like a griffin pointing the way to a guarded treasure.[29]

However, the caption claims it is *like* a griffin, not that it *is* a griffin and additional commentary also makes this distinction. In a video for the exhibition, Mayor has also backed off from her stronger claim in the book, and acknowledges that probably it will never be possible to know if the fossils were the origin for particular myths, rather than being explained within the context of a particular myth.

Whether the folklore of fossils will contribute significantly to our scientific understanding of fossils is an open question. Norell senses that most paleontologists probably do not think Protoceratops is actually the origin for the griffin myth. Furthermore, in terms of the kind of research that he is doing, which relies on the statistical analysis of large data sets, he agrees with Simpson. The study of ancient mythology is not going to be of much use to him. Nevertheless, he also still finds Mayor's ideas very interesting. He does not rule out the possibility that Mayor's ideas might be correct in regard to various other fossils. For example, it is possible to rearrange the bones of a mammoth skeleton into a two-legged giant. Moldavian peasants restored a rhinoceros in the configuration of a bipedal giant. Many fossil rhinoceros species have been found in Pikermi, Greece as well as other large fossil bones throughout the Mediterranean. It seems quite reasonable that the ancient Greeks might have interpreted these large bones as belonging to giants.

Norell also recognizes the importance of mounting an exhibition such as the Mythological Creatures to the museum. The museum has a three-pronged mission: research, education, and entertainment. You can't educate people if you don't first get them in the door and you can't get them in the door unless you entertain them. This exhibit was seen by more people than any other temporary exhibit in the entire museum's history including the very well received Darwin exhibit the year before.[30] It is next to the halls on fossils, which then provides museum goers an opportunity to really learn about the science of paleontology and evolution.

Even if Mayor's ideas in the long run do not prevail, in certain respects myths and science fulfill a similar function—they both provide humans beings with a presentation of the world and the forces that supposedly govern it. In Francois Jacob's words "They both fix the limits of what is considered as possible." When plesiosaurs along with other large fossils were first discovered they expanded the limits of what was possible. The meaning of fossils and the role they play not just in understanding the natural world, but also in understanding human civilization is being reassessed. And in this reassessment the boundaries between paleontology, anthropology, archeology, and history are shifting once again.

Evolutionary Psychology: Phrenology for the Twenty-first Century?

The role of nature in determining human nature is a question that has occupied humankind from the time our ancestors had brains big enough to contemplate such questions. Although numerous controversies have surrounded evolutionary theory from Darwin's time to the present, the ones that have garnered the most attention concern claims that have been made in regard to humans, specifically about our behavior. Although Darwin did not think humans were the exception to evolution, he was initially reluctant to elaborate how his theory applied to humans. He recognized that any discussion of human evolution would become emotionally charged, be a threat to religious beliefs, and highly politicized. This was true in Victorian times and remains so today. Evolution is about origins and our origins shape who we are today. In *Descent of Man* Darwin showed how his theory specifically applied to humans, including the evolution of a moral sense. Researchers have continued to build on Darwin's significant insights with varying degrees of success. The latest attempt to do so is evolutionary psychology.

The study of human evolution is particularly challenging for many reasons. The imperfection of the fossil record is exceptionally glaring in regard to human origins. Hominoid fossils are still relatively sparse. Dating of remains is often problematical and experts disagree as to whether certain fossils represent two different species, whether they should be categorized as two different genera, or that they merely represent sexual dimorphism within a single species. We cannot do experiments that recreate the past. Perhaps most important, behavior does not fossilize. The attempts to build an evolutionary theory of human nature dramatically illustrate the large role interpretation plays in determining the "facts" of human evolution. Facts only have meaning within a theoretical framework and the various theories that have been constructed have been loaded with the cultural values of the researchers. This is true of all science, but has especially plagued the field of human origins where gender, race, and nationalism have significantly influenced the construction of a theory. Because of these problems, it is not inappropriate to refer to theories about human origins as stories.[31] This is not meant in a pejorative sense or to downplay the immense amount of work that has been done in a variety of disciplines from molecular biology to archeology to primatology. It is merely meant to point out that a tremendous amount of interpretation exists in weaving research findings into a coherent theory about the origins of human behavior.

This story aspect of human origins is not something that can be gotten rid of by better methodology—that is by better quantitative analysis or higher standards for fieldwork. Improved methods never hurt any endeavor, but a major aspect of human evolutionary theory has been the actual construction of good stories. All stories are not equally good and the stories that are told are intimately linked with the questions that are asked, which in turn are a reflection of the sources of power, the intertwining of sex, race, and class of the researchers, and of the countries where the fieldwork is done.[32] A better story has to account more fully for what it means to be human or animal. It has to provide a more accurate description of humans and how we differ from apes and monkeys. But the question is what counts as a better, more coherent, fuller story?

The latest story that is being told is from the "new" field of evolutionary psychology. Many researchers claim that it is not really new, but rather warmed over sociobiology. It has a lot in common with phrenology as well. This, however, is not necessarily as damning as it might sound. Gall's ideas contributed enormously to our understanding of mind and behavior, even if phrenology eventually fell into disrepute. It remains to be seen whether evolutionary psychology will follow a similar path.

Many people maintain that what sets us off from our ancestors and our ape cousins is the development of culture. Certain evolutionary innovations were critical for that to happen. The ability to walk upright allowed the hands to be freed up to make tools. Tool use resulted in a feedback loop that led to a dramatic increase in our brain size that made possible the development of language. Our large brain is ultimately what makes possible all the rich cultural diversity that exists in the world today. However, our brain is not a cultural artifact; it is a biological organ. Evolutionary psychologists maintain that since we are still biological beings, anything that we do, even the many different forms of culture are ultimately rooted in our biology. Building on the ideas of sociobiology, they specifically want to use evolutionary theory to bridge the gap between biology and culture.

Evolutionary psychologists claim that the human mind consists of a set of cognitive mechanisms, discrete modules or "organs" that evolved as adaptations to the Pleistocene environment, referred to as the environment of evolutionary adaptation or EEA.[33] Just as the lung and kidney were shaped by natural selection to solve certain problems of physiology, the mind is a collection of mental organs shaped by natural selection to solve specific cognitive problems. That our brain is the product of evolution by natural selection is as reasonable an assertion as Gall's claim that the brain is the organ of the mind. However, just as with phrenology, many of the other claims of the evolutionary psychologists are highly questionable.

Evolutionary theory's greatest strength is unfortunately also its greatest weakness. Discovering the principle of natural selection to explain adaptation was a brilliant accomplishment, but it also resulted in a theory that was infinitely pliable. It is possible to make up an adaptive story about virtually any trait.[34] However, as Darwin observed, many traits are not adaptations at all. Instead, the process of development resulted in certain laws of correlation modifying traits independently of utility. Nevertheless, our large brain, which makes possible the complex behavior of Homo sapiens, must be regarded as the ultimate adaptation that has led us to being such an extraordinarily successful species. The problem is not regarding different behaviors as adaptations, but instead to evaluate what qualifies as a good adaptive story to explain a particular behavior.

Evolutionary psychology has attracted a lot of attention because it has claimed to provide explanations for highly interesting behaviors such as sexual jealousy, homicidal tendencies, and sexual attraction.[35] However, the interpretations of the evolutionary psychologists are problematic for a variety of reasons, but two points are particularly relevant in evaluating their claims. First, we do not have the kind of detailed knowledge of the Pleistocene environment necessary to illuminate the specific behaviors that humans would have evolved to solve specific problems. Other animals had to find food and mates, to take care of their young, avoid predation, and a variety of other tasks. If evolution has taught us anything, it is that organisms solve problems of survival in a myriad of ways, and that even within species tremendous variation in behavior exists. Second, assuming we can somehow manage to weed out all the biases of race, class, gender, and cultural superiority that continue to pervade the various evolutionary scenarios being offered about the origins of our behavior, a more fundamental problem underlies the whole discipline of evolutionary psychology. Just as with sociobiology, the explanatory power of evolutionary psychology depends on the belief that complex behavior is primarily genetically determined. Even if such an assumption is accepted (and as later discussion will show, this is a highly contentious issue), evolutionary psychologists face great difficulty in identifying the specific modules or clusters of traits that were specifically being selected for, let alone thinking that they will be primarily under the control of one or a few genes. In spite of the ridicule that phrenology received, Gall deserves great credit in attempting to define, organize and localize the different functions of the brain according to specific "organs." This has turned out to be extremely difficult and still has not been accomplished. Putting aside this more fundamental problem, is the adaptive story told by evolutionary psychologists to explain the differences between male and female behaviors a "good" story?

No one denies that differences in behavior exist between men and women, but exactly what are those differences and how much of the difference is rooted in biology? Evolutionary psychologists build their evolutionary tale of the war between the sexes on a set of five core assumptions: (1) Men are more promiscuous than women (2) while women are more interested in stable relationships. (3) Men are attracted to youth and beauty (4) while women are attracted to high-status men with resources as potential mates. (5) These preferences were shaped in the EEA and haven't changed significantly.

Assumptions about mate preferences are based primarily on surveys compiled by David Buss.[36] His surveys spanned the globe, as he collected data in thirty-seven different countries, from a wide array of cultures. From New York City to a village in Bangladesh, from New Zealand to China, from Paris to Buenos Aires, the surveys consistently revealed the same pattern. Men sowed their wild oats with some wild temptress, but when they were ready to settle down, they wanted a young, pretty, and virginal wife who would be faithful; and women wanted a mature, resourceful man. Even successful, financially secure women still wanted to marry men whose earning power and social status was equal to or preferably greater than their own. Females can't escape their innate preferences that were shaped in the EEA when they were tied down with an infant who they would have suckled for several years and they had to depend on their mate to bring home the mastodon.

Evolutionary psychologists predict that men will be most threatened by sexual infidelity while women will be most threatened by emotional infidelity.[37] Virtually all cultures have elaborate mechanisms in play to keep women faithful, and women who are found to be unfaithful are beaten, often killed, and made social outcasts. According to Margo Wilson and Martin Daly, male sexual jealousy is an adaptive solution to the problem of cuckoldry avoidance.[38] Unlike women, men cannot be absolutely sure that a child is their own and they do not want to be investing in unrelated offspring. Thus, they will go to great lengths to prevent that from happening and will be totally unforgiving if they find out that their mate has been unfaithful. Women certainly do not like that their partner sexually wanders; however, they will tolerate men's sexual infidelities, so long as they remain providers. They are much more threatened about an emotional attachment because this is more likely to result in a withdrawal of support, both financially and in caretaking of the children.

The story offered by the evolutionary psychologists appears to provide a good explanation of observed differences between male and female behavior. However, not only is there data that contradicts their story, a variety of other stories can be told that explain the differences in male and female behavior. We do not necessarily have to look deep into our evolutionary past to find another explanation as to why women want a man who makes a decent wage. Men make up about half the population, but they control between 75 percent and 95 percent of the world's wealth. In the United States a bachelor's degree adds $28,000 to a man's salary, but only $9,000 to a woman's. Women recognize, just as their sisters have throughout history, that the best and most efficient way to accumulate status and wealth is to marry it, rather than try to achieve it on their own.[39] Even women who are financially self-sufficient still prefer a mate of their own socioeconomic status. Professional women have found out that many men resent having a mate who is more successful than they are. It threatens their self-esteem and their status within the male hierarchy. Once women become convinced that men genuinely would be pleased to have their partner be high achievers, perhaps then they would stop being so concerned about a male's ability to earn a good living. Primatologist Sarah Baffer Hrdy wrote that "when female status and access to resources do not depend on her mate's status, women will likely use a range of criteria, not primarily or even necessarily prestige and wealth, for mate selection." A small number of professional women marry prisoners. The women are guaranteed sexual fidelity and are the beneficiaries of all their husbands' love and gratitude. Hrdy commented that "peculiar as it is, this vignette of sex-reversed claustration makes a serious point about just how little we know about female choice in breeding systems where male interests are not paramount and patrilines are not making the rules."[40] For example, could our female ancestor in the EEA of the Pleistocene tell the difference between a sexual dalliance and a serious threat to her pair bond that would leave her and her offspring destitute? Furthermore, we don't really know if our ancestors formed pair bonds, although that is a prevailing assumption in the primatology and paleoanthropology literature.

Evolutionary psychologists' theories about behavior differences between men and women are very close to Darwin's original theory of sexual selection. According to Darwin's theory of sexual selection, males compete for females and the females choose. Anything that would make the male more attractive to the female would increase his chances of getting his genes into the next generation, hence the spectacular plumage of the peacock while the peahen remained quite drab. Furthermore, the female is highly discriminating or coy, while the male would court and mate anything he possibly could. However, examining Darwin's idea of the coy female a little more closely presents serious problems. In spite of Darwin's brilliance and his willingness to go against cherished beliefs, he, nevertheless, could not escape his time. In many respects Darwin's theory reads as if he grafted Victorian morality onto the animal kingdom. In this scenario, females could always find someone to mate with them, and thus, they could afford to be choosy. Furthermore, there was very little variance in female reproductive success. However, because of male-male competition for the attentions of the coy female, their reproductive success varied greatly.

Evidence for this difference in variance between male and female reproductive success was confirmed by meticulous experiments conducted by Angus John Bateman in the *Drosophila* or fruit fly. While 21 percent of the males produced no offspring at all, only 4 percent of the females made such a poor showing. Furthermore, a successful male could produce nearly three times as many offspring as the most successful female. Bateman suggested that it would always be to the advantage of the male to mate just one more time. There would be strong selection pressure on males to be undiscriminating and eager to mate. Not only should females be highly discriminating in their choice of mates, but also they should be uninterested in mating more than once or twice, because they would already be breeding close to capacity, even after just one copulation. There would be selection pressure on them to pick the best and strongest male to father their offspring.

Robert Trivers, a leading evolutionary theorist and architect of sociobiology, built on the ideas of Bateman. There are two central themes to Triver's evolutionary story. First, the nurturing female who invests much more per offspring than the male is contrasted with the competitive male who invests little more beyond sperm, but actively competes for access to any additional females. Since the egg is so much bigger than sperm, it is energetically much more costly to make and there will be selection pressure on females to protect their eggs. Since female investment is already so large, it was assumed that it could not be increased and that females were already breeding close to capacity and there would be little variance in female reproductive success just as Bateman's experiments showed (Although one might ask why so much weight should be given to the behavior of fruit flies in building models of human behavior). Males, on the other hand, were making sperm all the time, at supposedly energetically little cost, and so they can disseminate them indiscriminately. A corollary to these ideas was implicit in many of Darwin's writings, but not explicitly stated: since females were always breeding to capacity, selection acted primarily on the males. Darwin had plenty of evidence that this was the case. One only had to look at the beautiful plumes of the male peacocks, the huge antlers of the male deer, and the generally bigger size of males to conclude that natural selection exerts its power on the male of the species.

This story seems quite reasonable. Clearly, males have the ability to inseminate multiple females while females are inseminated at most once during each breeding period. But just as with phrenology, contrary data is being ignored. In some species of fish, insects, and cats several fathers are involved in a single brood. In addition, not only was there no cost accounting of how much energy it takes to make all that sperm, but more importantly, the energetic costs of competing with other males for access to females was also ignored, whether it was growing bigger antlers or actually engaging in fights, which is quite common through the animal world.

A lot of data about female behavior has also been ignored. First, it is a mistake to assume that females are not competing among themselves for resources, status, territory, and a variety of other factors to help insure that they successfully raise their offspring to maturity. This has been well documented in several different species of primates.[41] Second, what about the females that forgot to be coy?[42] It has been known for years that although many bird species mate for life, many a cheater has been found amongst them. With the advances in DNA technology, it now is possible to unequivocally identify paternity. Many a female bird is less than chaste. Female cats are notoriously promiscuous. A lioness will mate up to one hundred times a day with many different males for the six to seven day period she is in estrous. Nonhuman primates engage in a wide range of mating behaviors. Female savanna baboons initiate multiple brief consortships while female chimpanzees alternate between prolonged consortships with one male and communal mating with all males. DNA studies of the Gombe chimpanzees, made famous by Jane Goodall, show that half of the offspring in a group were not the offspring of the resident males. These females left their local environment, escaping the scrutiny of their human observers, and became impregnated by outside males. They did this at great risk to themselves and their own offspring. Male chimps closely guard the movement of fertile females, threatening them, and they will sometimes kill an infant they do not think is their own. Bonobos, our closest relatives have a particularly wild sex life. They do it all the time and both males and females engage in frequent homosexual acts.

Finally, as many critics of the mate preference surveys have pointed out, women and men are much more similar than different in what they want in a prospective partner. In every culture, both women and men rate love, dependability, emotional stability, and a pleasant personality as the four most important traits in a mate. Only in the fifth tier do the differences that the evolutionary psychologists claim are so basic emerge.

In spite of these criticisms, females do get pregnant and have babies; males do not. It is unlikely that natural selection would not have produced different adaptations for this most fundamental difference. But just as phrenology reinforced prevailing stereotypes about different races, classes, and gender, evolutionary psychology seems to be following a similar path in perpetuating the prevailing stereotype of coy, passive females and competitive, promiscuous males. These ideas have been picked up and reinforced in the popular writings about sociobiology. Playboy had a feature article entitled "Darwin and the Double Standard." Cognitive psychologist Steven Pinker wrote in the *New Yorker* that President Bill Clinton was only acting out his evolutionary heritage with his numerous sexual affairs. "Most human drives have ancient Darwinian rationales. A prehistoric man who slept with fifty women could have sired fifty children, and would have been more likely to have descendants who inherited his tastes. A woman who slept with fifty men would have no more descendants than a woman who slept with one. Thus, men should seek quantity in sexual partners; women quality."[43]

Once again, a different evolutionary story could be told. Females take great risks to philander, but they do. Even if we accept the evolutionary psychologists' explanation of jealousy, it does not explain why female primates should be "promiscuous." There are energy costs to finding nonresident males to mate with. Females who wander risk losing the protection of one particular male, of being attacked by a jealous male, of having their offspring killed, and of contracting venereal disease by having multiple partners. To even begin to address these issues means recognizing the limitation of sexual selection theory.

Male-male competition and female choice would be only one aspect of a different evolutionary story. In addition, genetic benefits are postulated for offspring of mothers who were sexually assertive as well as nongenetic benefits for mothers and/or progeny. This story looks at the world from a female point of view. Hrdy suggests a variety of hypotheses as to why "promiscuous" behavior would be selected for in females. First, sperm from a number of males insures conception. If a female has only one partner who is sterile, her genes will be lost forever to future generations. In addition, variety is key to evolutionary success. More than one father helps ensure that offspring will have a variety of different traits, some of which will be adaptive in a constantly changing environment. In species where litters have more than one father, diverse paternity means that females may be able to enlist aid of all the males for all the offspring because the males do not know which one is their own. Finally, if a female is successful in soliciting copulation from a higher-ranking male, this could possibly result in stronger offspring. Various nongenetic benefits are postulated for several different hypotheses. Multiple matings are possibly physiologically beneficial in making conception more likely. In the "prostitution" hypothesis, females exchange sexual access for resources, enhanced status, and so on. The "keep 'em around" hypothesis suggests that with the approval of the dominant male, females solicit subordinate males to discourage them from leaving the group. Finally, the "manipulation hypothesis" claims that by confusing males about the paternity of the offspring females are able to extract investment in or tolerance for their infants from different males.

A more complex story can be told about male behavior as well. There are many benefits for a male who bonds with one female and invests heavily in her offspring whom he can be reasonably sure are his own. Data from nonhuman primates as well as cross cultural studies in human societies show that females do a great deal more than what is defined as traditional "mothering behavior" that influences the survival of their offspring. Males, far from being just sperm donors, also engage in a wide range of behavior that also affects the outcome of offspring. The stunning film *The March of the Penguins* beautifully documented the importance of the male penguin in the successful raising of offspring. Females are more political and males more nurturing than traditional sexual selection theory suggests. It is apparent that a variety of different strategies can result in reproductive success. Diversity and variety has always been and remains the key to evolutionary success.

Many different adaptive stories can be invented to explain the differences in behavior between men and women. Males and females have different interests and they will inevitably tell different stories. When women entered the field of primatology in great numbers in the 1960s many core assumptions were challenged. As data continues to be collected new stories will be proposed. Evolutionary theory has and will continue to contribute to our understanding about men and women, but sorting out a "good" adaptive story from a "bad" one remains extremely difficult. Because it concerns such highly emotionally charged issues that have all kinds of practical and political implications, there may never be consensus. Finally, even if the claim that our brain evolved as series of cognitive mechanisms, discrete modules or organs as adaptations to the Pleistocene environment is correct, we know very little about the EEA. Our knowledge is based on a few skulls, partial skeletons, and a handful of disputed artifacts, which makes possible the Kiplingesque "just so" story telling that has been discussed. However, this is not the only area where our knowledge is woefully lacking.

A more fundamental problem underlies the program of the evolutionary psychologists than sorting out adaptive stories and this is where the parallels to phrenology are especially relevant. In a research tradition that traces its lineage back to Gall, faculty psychologists argue that mental behavior is the result of several distinct psychological mechanisms.[44] Although many cognitive psychologists agree that the brain is organized into distinct modules or organs just as Gall stated (even if it is recognized that these functional modules are diffusely located throughout the brain), there is very little consensus on exactly what those modules are beyond general ones such as a language system and a perceptual system. In spite of this, evolutionary psychology has received considerable attention. At a popular level its explanation of interesting behaviors, particularly differences between men and women insured that it would find a following. However, its scientific legitimacy depends on its appeal to genetics.

The twentieth and twenty-first centuries are a testimonial to the spectacular successes of molecular biology that have infused popular culture, influencing beliefs, attitudes, and expectations. Designer genes, the new reproductive technologies, cloning, and the complete sequencing of the human genome, all are testimony to the power this knowledge has to shape our lives. The elucidation of the structure of DNA ushered in a new era of trying to understand human character in terms of biology. Research in neurology, genetics, and physiology provide overwhelming evidence of the enormous role biology plays in who we are as individuals. Our understanding of neurobiology is increasingly sophisticated and fine grained, revealing an intricacy far beyond anything Gall could have even imagined. The overwhelming message from this research is of complexity and interaction. For example, research on the personality trait "novelty seeking," which many psychologists think is one of four basic building blocks of normal temperament, illustrates the difficulty in defining a trait and elucidating the genetic contribution to such a trait.[45]

Researchers think that genetics contributes about 50 percent to a particular character trait. Two independent studies looked at variants of a gene that codes for the D4 dopamine receptor (D4DR), one of five known receptors that play a role in the brain's response to dopamine, and correlated those findings with the results of personality questionnaires.[46] People who test high for this trait are extroverted, impulsive, extravagant, and exploratory. Among the known neurotransmitters, dopamine is the most strongly linked to pleasure and sensation seeking.[47] Assuming that each gene has more or less the same influence as the D4 receptor, this suggests that there are four or five genes involved with the trait. But scientists readily admit that other genes exist that haven't yet been identified. At best, the dopamine receptor accounts for perhaps 10 percent of the variance in novelty seeking behavior in the population as a whole. Nevertheless, this was touted as a major finding about genes and personality. The *New York Times'* caption "Variant Gene Tied to a Love of New Thrills" belies the difficulty in studying behavior as well as the restricted nature of the results.[48] The novelty seeking gene results have failed to be replicated. In a different study, in one of the cohorts, researchers found a significant association in the opposite direction as previously observed, suggesting that the role D4DR plays in personality might need to be reevaluated.[49] This research, in spite of being meticulously done, made certain assumptions that were reasonable, but just like Gall's assumptions, they may be wrong. In addition, researchers do not agree whether there are four basic building blocks of temperament, whether personality tests accurately measure the different traits and whether such traits are primarily under genetic control.[50]

Trying to elucidate the biological components of behavior is not misguided, but careful researchers emphasize both the limited nature of a particular finding and the complexity of the phenomenon they are studying. "The power of genes is real but limited—a principle that operates even during the growth of the embryo. . . . No human quality, physical or psychological, is free of the contribution of events within and outside the organism. Development is a cooperative mission and no behavior is a first-order, direct product of genes."[51] While an inherited neurochemical and physiological profile that is linked to emotion and behavior is part of our modern-day definition of temperament, the relationship between physiology and emotion and behavior is complex and not yet understood. It is a mistake to assume that measurable differences in biological traits such as enzyme levels, brain structures, and receptor proteins in neurons is prima facie genetic. Any learned behavioral response has a biological component, but it is not genetic in the sense of a transmissible element. Furthermore, a person's subjective feelings do not map well onto the presumably relevant physiological indexes.[52] In spite of such poor linkage between physiological and psychological states, evolutionary psychologists claim that mental faculties are evolutionary adaptations resulting from gene action in ontogeny. Finally, even if the research ultimately vindicates the dopamine receptor gene as important in "novelty seeking," it is doubtful that any cognitive psychologist would claim there is a module for novelty seeking. For the moment, evolutionary psychology has limited application because of our current knowledge in cognitive science.[53] By using the language of genetics and evolution, evolutionary psychologists mask the fact that there is little direct experimental evidence in favor of their theory.

Nevertheless, biological theories of behavior continue to capture the imagination of scientists and the general public in spite of a long and ignominious history culminating with the eugenic policies of the Nazis. Biological arguments generally have been used to discriminate and oppress people.[54] Biology is destiny, or so the argument goes. However, as with phrenology, this analysis misses an important aspect of why biological theories of behavior are so seductive. One might be born with a certain set of lumps and bumps of the skull, but phrenologists maintained that it was possible through testing, and hard work to develop those organs that were beneficial, and suppress ones that were not. With the dawn of the new millennium, the human genome project offers us the same possibilities. Gall's faculties or organs have metamorphosed into genes. We can screen for potential devastating diseases such as cancer and diabetes, identify risk factors, and live our lives accordingly. The genome project promises to place within our power the ability to change our character. Not only are there genes for fatness, but there are genes for shyness, aggressiveness, and novelty seeking as well. The gene prophets, just like the phrenologists, suggest that once we fully understand human nature at the molecular level, it will be possible to screen for the whole panoply of traits that make up human nature whether it be loyalty, or shiftlessness, honesty or criminality, genius or stupidity. Armed with such knowledge, we will finally be able to rid society of problems that so far appear to be intractable. The great appeal of the genome project is that biology is *not* destiny; it is malleable.[55] We *can* change our nature. Thus, ironically, while evolutionary psychology is claiming that our uniquely human nature was shaped back in the Stone Age and can't be changed, the plausibility of its arguments rests on the conviction that behavior is primarily the result of genetics, which now can be manipulated.

Evolutionary psychologists, bolstered by the human genome project, promise to locate mental attributes in our genes. We correctly ridicule the conclusions drawn from the simple craniology that phrenology degenerated into. However, as we read about genes for criminality or alcoholism, or for novelty seeking, one can't help wondering how different this is from the phrenologist's claim that distinct organs in the brain were responsible for traits such as secretiveness, destructiveness, and spirituality. Nevertheless, like phrenology, it would be a mistake to dismiss evolutionary psychology as mere pseudoscience. Gall's fundamentally functional approach that claimed anatomical and physiological characteristics directly influence mental behavior continues to be the basis for our present-day research on mind and brain. Evaluating the stories that evolution tells will always be difficult; nevertheless, natural selection and the concept of adaptation are extremely powerful concepts in trying to understand human nature.

Evolutionary psychology is an example of frontier science that has many parallels to phrenology with its mixture of important insights and claims that have little experimental evidence. Mainstream neurobiological research absorbed the ideas from phrenology that provided a viable research program. The "bumpology" was discarded, and it was discarded because the quality of the evidence in support of it was so poor. However, what counts as "poor" or "good" evidence is continually reevaluated. Particularly in the study of human nature, culture plays an enormous role in making that kind of judgment. The meaning of the terms nature and culture have been heavily contested throughout history with the idea of biological determinism going in and out of vogue. The evolutionary psychologists are telling a story about why humans act the way they do. A variety of factors from the power and prestige of molecular biology to cultural attitudes about males and females influence both the construction and evaluation of the stories. Like its close relation sociobiology, it has the potential to contribute enormously to our understanding of human behavior, but there are significant problems that must be overcome before it will be able to do so. It is negotiating for a place at the high table of science, but may instead end up at the counter of pseudoscience.

A Few Concluding Remarks

Evolution, like history, is about interpreting the past. Also like history, understanding the past can provide insight into the present and the future. But the past is gone; we cannot recreate evolutionary history. We must rely on different kinds of evidence to provide clues to help us reconstruct the past. The evidence is of many different kinds, from molecular sequence data to fossil bones. Evolution draws on findings from geology, psychology, anthropology, archeology, zoology, physiology, anatomy, and genetics to tell a coherent story about life's history.[56] This storytelling aspect pervades all areas of evolutionary theorizing, from origin stories about the first forms of life to why the dinosaurs went extinct to the prevailing patterns of evolutionary change to the emergence of consciousness. The stories that are told are continually modified, and many are discarded. This is the essence of the scientific process. Evidence is often contradictory and new technologies in their early stages are not necessarily better in supplying reliable information. For example, DNA sequence analysis has revolutionized the classification of organisms, particularly in the bacterial world.[57] The technology has been extended to fossils. Although it is now possible to extract DNA from fossilized bones, a major problem has been that the samples often become contaminated with present-day bacteria or other mycological material. When Allan Wilson presented the first reliable DNA-based data on the relationship of mammoths to modern-day organisms, he quipped, "We have learned today that the mammoth was either an elephant or a fungus."[58] What this means is that science is fraught with error. However, error does not fundamentally undermine the legitimacy of scientific knowledge. Rather, past error is a form of negative knowledge and can be used to guide further research.[59]

Science is as successful as it is because it has developed a set of standards and a methodology for designing experiments, interpreting results, and constructing effective scientific institutions. This does not prevent scientists from making mistakes, but the various aspects of scientific practice mean that science has enormous capacity to be self-correcting. Nevertheless, since scientists are only human, science can never become totally objective. However, objectivity is a value that all scientists subscribe to, and as a value it means that scientists sincerely believe that nature plays an enormous role in the acceptance or rejection of scientific claims. Certainly, scientists compete for prestige, power, and recognition. Some may be immoral and unscrupulous human beings. Some clearly use their scientific expertise to push political or social agendas when the evidence clearly does not support such a position, therefore crossing over the boundary that separates science from pseudoscience. Furthermore, scientists play a powerful role in delineating those boundaries. Nevertheless, such practices do not seriously weaken the effectiveness of science, and this is because science is an investigation of the natural world. Nature may be tricky, but ultimately does not lie. Science has moved to a more and more accurate representation of how nature really is. It has been able to do so because of the standards of evidence. Experiments need to be repeatable. If it is not possible to do experiments such as the case with many aspects of evolutionary theory, then a variety of different kinds of evidence need to be gathered and interpreted. Nature plays a critical role in determining what eventually becomes reliable knowledge. It is for this reason that the examples of marginal science in this book remained marginal. The practice of science, like any human endeavor, is imperfect. Science tells an ever-changing story about the nature of the material world. But, unlike a fictional story, whose author is free to manipulate it in any way she chooses, scientists are heavily constrained in the stories they tell. Nature is a strict taskmaster, but her strictness ensures that scientific practice is the most powerful tool we have for generating knowledge about the universe. Yet it must always be remembered that an essential part of that practice is that it is a collective human activity, embedded within society as a whole. It is this fundamentally human aspect that makes possible the richness and diversity of what we call science.

Notes

Preface

1. David Knight, *The Age of Science* (New York: Blackwell, 1986).

1. Introduction

1. David Knight, *The Age of Science* (New York: Blackwell, 1986).

2. Thomas Huxley, "March 22, 1861," *Life and Letters of Thomas Huxley*, 2 vols., ed. Leonard Huxley (New York: Appleton, 1900), 1:205.

3. Walter Houghton, *The Victorian Frame of Mind, 1830–1870* (1957; repr., New Haven: Yale University Press, 1985), 1; but see the entire first chapter about the kinds of changes taking place in Victorian society.

4. Herbert Spencer, *Social Statics: Or the Conditions Essential to Human Happiness Specified and the First of Them Developed* (London: Chapman, 1851).

5. See Bernard Lightman, ed., *Victorian Science in Context* (Chicago: University of Chicago Press, 1997).

6. Simon Knell, *The Culture of English Geology, 1815–1851* (Burlington, VT: Ashgate, 2000).

7. Robert Young, *Mind, Brain and Adaptation in the Nineteenth Century* (Oxford, UK: Clarendon Press, 1970).

8. (Chambers, Robert), *Vestiges of the Natural History of Creation* (1844; repr., Surrey, UK: Leicester University Press, 1969).

9. Charles Darwin, *The Descent of Man and Selection in Relation to Sex* (London: John Murray, 1871; Princeton, NJ: Princeton University Press, 1981).

10. John Morley, *Recollections*, discussing reviews of Darwin's *The Descent of Man*, 1872, quoted in Houghton, *The Victorian Frame of Mind*, 59.

11. This section has been heavily informed by John H. Brooke's *Science and Religion: Some Historical Perspectives* (Cambridge: Cambridge University Press, 1991).

12. William Paley, *Natural Theology* (London: Farnbourgh, Gregg, 1802; repr., 1970).

13. Thomas Paine, *The Age of Reason: Being an Investigation of True and Fabulous Theology* (Paris: Barrois, 1794), 36-37.

14. Ibid., 46.

15. Brooke, *Science and Religion.*

16. See also Herbert Spencer, *Principles of Psychology: A System of Synthetic Philosophy*, 2nd ed., 2 vols. (London: Williams and Norgate, 1872).

17. Charles Darwin, *The Origin of Species* (London: John Murray, 1859; New York: Avenel Books, 1979), 459.

18. Alison Winter, "The Construction of Orthodoxies and Heterodoxies in the Early Victorian Life Sciences," in Lightman, *Victorian Science in Context*, 24–50.

19. Adrian Desmond, *The Politics of Evolution: Morphology, Medicine, and Reform in Radical London* (Chicago: University of Chicago Press, 1989).

20. There is an enormous literature on this topic, but a few relevant citings include: M. P. Hanen, M. Osler, and R. G. Weyant, eds., *Science, Pseudo-Science, and Society* (Waterloo, ON: Wilfrid Laurier University Press, 1980); Thomas Kuhn, *The Structure of Scientific Revolutions* (Chicago: University of Chicago Press, 1962; repr., 1970); Rachel Lauden, ed., *The Demarcation between Science and Pseudo-Science*, vol. 2 (Blacksburg: Center for the Study of Science in Society, Virginia Polytechnic Institute, 1983); Seymour Mauskopf, "Marginal Science," in *Companion to the History of Modern Science,* ed. R. C. Olby, G. N. Cantor, J. R. R. Christie, and M. J. S. Hodge (London: Routledge, 1990), 869–885; Paolo Parrini, Wesley C. Salmon, and Merrilee H. Salmon, eds., *Logical Empiricism—Historical and Contemporary Perspectives* (Pittsburgh: University of Pittsburgh Press, 2003); and Karl Popper, *The Logic of Scientific Discovery* (London: Routledge Classics, 1959; repr., 2002).

21. Henry Bauer, *Scientific Literacy and the Myth of the Scientific Method* (Chicago: University of Illinois Press, 1992).

2. Swimming at the Edges of Scientific Respectability

1. William Crafts, *The Sea Serpent or Gloucester Hoax: A Dramatic Jeu d'Espirit in Three Acts* (Charleston, SC: A. E. Miller, 1819), 16.

2. See Martin J. S. Rudwick, *Scenes from Deep Time: Early Pictorial Representations of the Prehistoric World* (Chicago: University of Chicago Press, 1992), 42–43; and Richard Ellis, *Monsters of the Sea* (New York: Knopf, 1994), 23.

3. Earth history was divided into three main periods: Tertiary or Cenozoic, Secondary or Mesozoic and, the oldest, Primary or Paleozoic.

4. Heinrich Rathke, quoted in Bernard Heuvelmans, *In the Wake of the Sea-Serpents,* trans. Richard Garnett (New York: Hill and Wang, 1968), 48. Heuvelmans provides the most thorough account of the sea serpent's history.

5. William J. Hooker, quoted in Heuvelmans, *In the Wake,* 24.

6. John Gordon, *A Collection of Scientific Articlesand Newspaper Reports from 1816–1905 Relating to the Sea Serpent* (South Brewer, ME: John Gordon, 1926), 7, n. 1.

7. George M. Eberhart, *Monsters* (New York: Garland, 1983). Eberhart's list on sea monsters runs thirty-eight pages and includes only articles in English and omits all newspaper articles, travelogues and tales of "weird creatures."

8. Quoted in Heuvelmans, *In the Wake*, 48.

9. Ibid., 49.

10. Quoted in Ellis, *Monsters of the Sea*, 41-42.

11. Ibid., 44.

12. Quoted in Heuvelmans, *In the Wake*, 50.

13. Quoted in Ellis, *Monsters of the Sea*, 45.

14. Quoted in Stephen Gould, *Bully for Brontosaurus* (New York: Norton, 1991), 271.

15. For further discussion of these points, see Martin Rudwick, "Uniformity and Progression: Reflections on the Structure of Geological Theory in the Age of Lyell," in *Perspectives in History of Science and Technology,* ed. Duane H. Roller (Norman: University of Oklahoma Press, 1971), 214; and Peter Bowler, *Fossils and Progress: Paleontology and the Idea of Progressive Development in the Nineteenth Century* (New York: Science History Publications, 1976).

16. A great deal of secondary literature on uniformitarianism exists. A few relevant citings are Peter Bowler, *Evolution* (Berkeley and Los Angeles: University of California Press, 1984); Walter Cannon, "The Uniformitarian-Catastrophist Debate," *Isis* 51 (1960): 38–55; Charles Gillispie, *Genesis and Geology: A Study in the Relations of Scientific Thought, Natural Theology and Social Opinions in Great Britain 1790-1850*, (1951; repr., New York: Harper, 1959); Stephen Gould, "The Eternal Metaphors of Paleontology," in *Patterns of Evolution,* ed. A. Hallam (Amsterdam: Elsevier, 1977), 1–26; Stephen Gould, "Agassiz's Marginalia in Lyell's *Principles* or the Perils of Uniformity and the

Ambiguity of Heroes," *Studies in the History of Biology* 3 (1979): 119-138; R. Hooykaas, *Natural Law and Divine Miracle: The Principle of Uniformity, Biology, and History* (Leiden: Brill, 1959); and Martin Rudwick, "Uniformity and Progression."

17. James Hutton, *Theory of the Earth, Transactions Royal Society of Edinburgh* 1 (1788), 304.

18. Archives, Boston Museum of Science.

19. Quoted in Heuvelmans, *In the Wake*, 151.

20. "The Sea Serpent," *Niagara Patriot* 1:18 (August 18, 1818), 2, c. 4–5.

21. *Niagara Patriot* 1:20 (September 1, 1818), 2, c. 4.

22. "The Serpent Not Taken," *Niagara Patriot* 1:23 (September 22, 1818), 2, c. 3–4.

23. "The Sea Serpent," *Buffalo Patriot* 5:225 (August 6, 1822), 3, 4, reproduced from the *Boston Gazette.*

24. "Sea Devil!" *Buffalo Patriot* (June 3, 1823), 3, c. 3.

25. "Enormous Fish," *Buffalo Patriot* (October 7, 1823), 2, c. 4.

26. The group of aquatic mammals that includes whales, porpoises and dolphins.

27. J. Dawson to C. Lyell, September 20, 1845, Charles Lyell Scientific Correspondence, Edinburgh University Library, Box 26.

28. See Heuvelmans, *In the Wake,* ch. 5.

29. Charles Lyell, *A Second Visit to the United States of North America*, 2 vols. (London: John Murray, 1847), 1:113.

30. From the *Illustrated London News*, October 28, 1848, Charles Lyell Scientific Correspondence, Edinburgh University Library, Box 26.

31. Nicolaas Rupke, *Richard Owen: Victorian Naturalist* (New Haven, CT: Yale University Press, 1994), 324–332.

32. Richard Owen, "The Great Sea Serpent,"*The Annals and Magazine of Natural History*, 2d ser., 2 (1848): 458–463.

33. Lyell to G. Mantell, February 20, 1830, in Charles Lyell, *Life and Letters of C. Lyell*, 2 vols., ed. K. Lyell (London: John Murray, 1881), 1:262.

34. Lyell, *A Second Visit,* 114.

35. Dov Ospovot, "Lyell's Theory of Climate," *Journal of the History of Biology* 10 (1977): 317–339.

36. Georges Cuvier, 1823, quoted in Heuvelmans, *In the Wake*, 182.

37. These numbers are taken from Heuvelmans, *In the Wake*, 35.

38. Carolyn Kirdahy, archivist, Boston Science Museum, personal communication, June 3, 1993.

39. Charles Cogswell, "The Great Sea-Serpent," *Zoologist* 6 (1848): 2316-2323.

40. Quoted in Harriet Ritvo, "Professional Scientists and Amateur Mermaids: Beating the Bounds in Nineteenth-Century Britain," *Victorian Literature and Culture* 19 (1991): 287.

41. Quoted in Ellis, *Monsters of the Sea*, 60.

42. Knell, *The Culture of English Geology* (ch. 1, n. 5,), xi–xviii.

43. Lyell to Whewell, September 30, 1840, quoted in Jack Morrell and Arnold Thackray, *Gentlemen of Science* (Oxford, UK: Clarendon Press, 1981), 117.

44. Morrell and Arnold, *Gentlemen of Science*, 32.

45. William Whewell, 1833 *BAAS Report*, xx–xxiv, quoted in Morrell and Arnold, *Gentlemen of Science,* 270.

46. Ibid.

47. Edward Newman, "The Great Sea Serpent," *Westminster Review* 50 (1849): 492.

48. Charles Lyell, *Principles of Geology Being an Inquiry How Far the Former Changes of the Earth's Surface Are Referable to Causes Now in Operation,* 3 vols. (London: John Murray, 1830–1833; repr., Chicago: University of Chicago Press, 1990), 1:123.

49. See Rudwick, *Scenes from Deep Time,* 48–49.

50. Rupke, *Richard Owen,* 330.

51. Ron Westrum, "Knowledge about Sea Serpents," in *On the Margins of Science: The Social Construction of Rejected Knowledge*, Sociological Review Monograph 27, ed. Roy Wallis (Keele, UK: University of Keele, 1979), 296.

52. Heuvelmans, *In the Wake,* 204.

53. Westrum, "Knowledge about Sea Serpents," 298–300.

54. T. H. Perkins, letter to father, October 20, 1820, Charles Lyell Scientific Correspondence, Edinburgh University Library, Box 26.

55. Philip Henry Gosse, *Romance of Natural History* (London, 1860), 310–311.

56. Martin J. Rudwick, *The Meaning of Fossils* (New York: Neale Watson Academic Publishers, 1972; Chicago: University of Chicago Press, 1976).

57. Thomas Huxley, 1893, quoted in Heuvelmans, *In the Wake,*. 25. The Cretaceous was the most recent epoch of the Mesozoic period.

58. Bean and Goode, quoted in Heuvelmans, *In the Wake,* 25.

59. For a brief discussion of the history of the Dana eel larva, see Jorgen Nielsen and V. Larsen, "Remarks on the Identity of the Giant Dana Eel-Larva," *Vidensk. Meddr dansk naturh. Foren* 133 (1970):149–157.

60. Quoted in the *Ekstrabladet,* July 8, 1941.

61. Torben Wolff, August 25, 1993, personal communication. I wish to thank Dr. Wolff for informing me about the Dana eel larva. In addition

to sending me the article by Nielson and Larson, he also sent several Danish newspaper clippings and translated the relevant passages.

62. Jules Verne, *The Complete Twenty Thousand Leagues Under the Sea*, trans., intro., and notes Emanuel J. Mickel (1870; repr., Bloomington: Indiana University Press, 1969), 1.

3. Franz Gall, Johann Spurzheim, George Combe, and Phrenology

1. Marvin Gardner, *Fads and Fallacies in the Name of Science* (NewYork: Putnam's, 1952).

2. Robert Young, *Mind, Brain and Adaptation in the Nineteenth Century* (Oxford, UK: Clarendon Press, 1970).

3. George Combe (1788–1858) was the son of an Edinburgh brewer. He studied law at Edinburgh University and obtained his commission as a writer to the signet. He became arguably the foremost proselytizer of phrenology in Victorian England. His role in phrenology's history will be discussed later in the chapter.

4. *Weekly Medico-chirugical Review and Philosophical Magazine,* February 22, 1823, 34.

5. Franz Gall, quoted in Owsei Temkin, "Gall and the Phrenological Movement," *Bulletin of the History of Medicine* 21 (1947): 277–278.

6. Richard Chenevix, "Gall and Spruzheim: Phrenology," 1828, quoted in Roger Cooter, *The Cultural Meaning of Popular Science* (New York: Cambridge University Press, 1984), 5.

7. The exact origin of the actual term remains obscure. Gall and Spurzheim described an article by Thomas Forster in 1815 on the anatomy of the nervous system that used the term phrenology. Possibly it was used as early as 1805 by Benjamin Rush in some lectures when he mentioned "the state of phrenology if I may be allowed to coin a word to designate the science of the mind." See P. Noel and E. Carlson, "Origins of the Word 'Phrenology,'" *American Journal of Psychiatry* 127 (1970): 694–697.

8. Franz Gall, *On the Function of the Brain and of Each of Its Parts,* vol. 4:247, 260, quoted in Temkin, "Gall and the Phrenological Movement," 304.

9. Gall, quoted in Temkin, "Gall and the Phrenological Movement," 277–278.

10. Franz Gall and Johann Spurzheim, *Anatomie et physiologie du systême nerveux en général, et du cerveau en particulier, avec des observations sur las possibilitié de reconnoitre plusieurs despositions intelluelles et morales de l'homme et des animaux par las configuration de leurs têtes,* quoted in Tempkin, "Gall and the Phrenological Movement," 282.

11. Descartes' legacy continued to hamper research in neurobiology. See Antonio Damasio, *Descartes' Error* (New York: Avon Books, 1994).

12. Gall and Spurzheim, *Anatomie et Physiologie* (1810–1819), 2 (1812): 5, quoted in E. Clarke, and L. S. Jacyna, *Nineteenth-Century Origins of Neuroscientific Concepts* (Berkeley and Los Angeles: University of California Press, 1987), 226.

13. (Robert Chambers), *Vestiges of the Natural History of Creation* (1844; repr., Surrey, UK: Leicester University Press, 1969), 341. See also James Secord, *Victorian Sensation* (Chicago: University of Chicago Press, 2000), 85–87.

14. Young, *Mind, Brain and Adaptation in the Nineteenth Century.*

15. Ibid., 43.

16. "Charles Darwin, the Eminent Naturalist," *American Phrenological Journal and Life Illustrated* 48:4 (October 1868): 121–122.

17. See Robert Richards, *Darwin and the Emergence of Evolutionary Theories of Mind and Behavior* (Chicago: University of Chicago Press, 1987, ch. 3.

18. Charles Darwin to Joseph Hooker, 1844? in *Life and Letters of Charles Darwin,* 2 vols., ed. Francis Darwin (London: John Murray, 1887), 1:301-302. Hereafter cited as LLCD.

19. (Chambers), *Vestiges,* 335–336.

20. Darwin, *The Descent of Man,* 105.

21. (Chambers), *Vestiges,* 336.

22. Darwin, *The Descent of Man,* 42.

23. (Chambers), *Vestiges,* 336.

24. Darwin, *The Descent of Man,* 48–49, 77.

25. (Chambers), *Vestiges,* 337.

26. Darwin, *The Descent of Man,* 50.

27. (Chambers), *Vestiges,* 341–345.

28. Darwin, *The Descent of Man,* 52.

29. Ibid., 68.

30. Darwin, 1838, M notebook, 57, in Paul Barrett, ed., *Metaphysics, Materials & the Evolution of Mind* (Chicago: University of Chicago Press, 1974), 16.

31. Desmond, *The Politics of Evolution.*

32. See chapter 6 for a further discussion of evolution.

33. *Edinburgh Review,* 1815, 227.

34. Alfred Wallace, "The Neglect of Phrenology," *The Wonderful Century* (London: Swan Sonnenschein, 1898), 164–165.

35. George Combe, *The Constitution of Man* (1828; repr., Boston, MA: Marsh, Capen, & Lyon, 5th ed., 1835).

36. Secord, *Victorian Sensation,* 73.

37. *Lancet,* April 16, 1825, 41.

38. J. Fletcher, *Rudiments of Physiology,* ed. Robert Lewins (Edinburgh: John Carfrad and Son, 1837), 137–142.

39. Alison Winter, *Mesmerized* (Chicago: University of Chicago Press, 1998), 118.

40. Ibid., 117.

41. See Steven Shapin, "The Politics of Observation: Cerebral Anatomy and Social Interests in the Edinburgh Phrenology Disputes," in *On the Margins of Science: The Social Construction of Rejected Knowledge,* ed. R. Wallis (Keele, UK: University of Keele, 1979); and Cooter, *The Cultural Meaning of Popular Science* for a more complete discussion of this aspect of the debates over phrenology.

42. Review of Spurzheim's recently translated unpublished manuscript *Anatomy of the Brain, London Medical and Physical Journal* 56 (1826): 364–368.

43. Geoffrey N. Cantor, "The Edinburgh Phrenology Debate: 1802-1828," *Annals of Science* 32 (1975): 217.

44. In New Lanark, three thousand people lived in well-constructed sanitary houses and had a store where prices were low. Schools were provided for the workers' children. Wholesome recreation activities were provided and working conditions at the cotton mill that Owen owned were far better than what existed in the rest of England. New Lanark epitomized Utopian socialism.

45. Morton was considered a great data collector and epitomized the idea of the objective empirical scientist, but his analysis of the data has been severely criticized. See Stephen Gould, *The Mismeasure of Man* (NewYork: Norton, 1982), ch. 2.

46. Steven Shapin, "Phrenological Knowledge and the Social Structure of Early Nineteenth-Century," *Edinburgh Annals of Science* 32 (1975): 195–218, 226, n. 20.

47. "The Phrenological Organ of Tune," *The Art of Improving the Voice and Ear* (London: S. R. Bentley, 1825), cited in Roger Cooter, *Phrenology in the British Isles* (Metuchen, NJ: Scarecrow Press, 1989).

48. Winter, *Mesmerized*, 345–348.

49. Cantor, "The Edinburgh Phrenology Debate," 217.

50. This is not quite true. In Europe, Gall is much better known and held in much higher esteem. Further research is warranted in examining the history of phrenology in Europe.

51. Cantor, "The Edinburgh Phrenology Debate," 216–217.

52. See James Secord, ed., "Introduction," *The Vestiges of Creation and Other Evolutionary Writings* (Robert Chambers) (Chicago: University of Chicago Press, 1994), ix–xlv.

53. Secord, *Victorian Sensation*, 73–74.

54. Cantor, "The Edinburgh Phrenology Debate," 214.

4. The Crisis in Faith

1. David Strauss, quoted in John H. Brooke, *Science and Religion*, 271.

2. For a full discussion of the Buffalo physicians' investigations, see Vernon Bullough, "Spirit Rapping Unmasked: An 1851 Investigation and Its Aftermath," *The Skeptical Inquirer* 10 (1985): 61–67.

3. In automatic writing the medium, while in a trance, might find her hands writing independently of her own volition or be presiding when messages mysteriously appeared on slates. Even supposedly ordinary people often found their hands writing automatically, beheld visions, caused tables and furniture to gyrate as found in memoirs and diaries. See Janet Oppenheim, *The Other World* (Cambridge: Cambridge University Press, 1985), 9. Spirit photographers produced photographs of deceased relatives and other supernatural phenomenon. Oppenheim's book remains the best treatment of spiritualism and psychic research in the Victorian period.

4. Ibid., 8.

5. J. J. Thomson, *Recollections and Reflections*, 153–158, quoted in Oppenheim, *The Other World,* 339-340.

6. Quoted in Trevor Hall, *The Engima of Daniel Home* (Buffalo, NY: Prometheus Books, 1984), 34.

7. William Crookes, "Spiritualism Viewed by the Light of Modern Science," *Quarterly Journal of Science,* July 1870; also quoted in Oppenheim, *The Other World*, 344.

8. Quoted in Hall, *The Engima,* 48.

9. See Gordon Stein, *The Sorcerer of Kings* (Amherst, NY: Prometheus Books, 1993), 88–91.

10. For a full account surrounding Mrs. Lyon, see Hall, *The Enigma,* 88–93.

11. E. B. Tylor, "Ethnology and Spiritualism," *Nature* 5 (February 29, 1872): 343, in Oppenheim, *The Other World*, 16.

12. Douglas Daniel Home, *Lights and Shadows of Spiritualism* (London: Virtue, 1877), 326.

13. Sergeant Cox to Home, March 8, 1876, in Home, *Lights and Shadows of Spiritualism*, 326–327.

14. Oppenheim, *The Other World*, 340–343. Crookes had made these claims in articles he had written to the *Spiritualist Newspaper.*

15. Trevor Hall, *The Spiritualists*, 1962; repr. as *The Medium and The Scientist* (Buffalo, NY: Prometheus Books, 1984).

16. Ibid., 27–30.

17. Sergeant Cox, *Spiritualist Newspaper*, May 15, 1874, 230, quoted in Hall, *The Medium and the Scientist,* 76. It is worth noting that at first the *Spiritualist Newspaper* refused to publish this account and Cox

wrote a letter to *The Medium Daybreak,* explaining what had happened. Both periodicals finally published his account.

18. Letter to the *Spiritualist Newspaper*, June 5, 1874, quoted in Oppenheim, *The Other World*, 341.

19. In addition to Hall's book, see also the exchange between Hall and Ian Stevenson in three issues of the *Journal of the American Society for Psychical Research* 57:4 (1963): 215–226; 58:1 (1964): 57–65; 58: 2 (1964): 128–133.

20. Crookes, *The Spiritualist,* June 5, 1874, quoted in Hall, *The Medium and the Scientist,* 86.

21. Ibid., 96, n. 3.

22. Crookes's Society for Psychical Research presidential address, quoted in Oppenheim, *The Other World*, 350.

23. Balfour Stewart to William Barrett, December 26, 1881, quoted in Oppenheim, *The Other World*, 330.

24. Darwin, *The Origin*, 458.

5. Morals and Materialism

1. Charles Darwin, *The Autobiography of Charles Darwin and Selected Letters,* ed. Francis Darwin (New York: Dover, 1958), 9.

2. Lyell, *Principles*, vol. 2.

3. Darwin, *The Autobiography*, 42.

4. Ibid.

5. Alfred Russel Wallace, *My Life: A Record of Events and Opinions*, 2 vols. (London: Chapman and Hall, 1905), 1:87.

6. Ibid., 88.

7. bid., 194–195.

8. Ibid., 235.

9. Quoted in H. Lewis McKinney, *Wallace and Natural Selection* (New Haven, CT: Yale University Press, 1972), 6.

10. Alfred Russel Wallace, *The Malay Archipelago* (1869; repr., New York: Dover, 1962), 456.

11. Darwin to Charles Lyell, June 18, 1858, in LLCD, 1:473.

12. Charles Darwin, December 22, 1857, quoted in James Marchant, *Alfred Russel Wallace: Letters and Reminiscences* (New York: Harper Brothers, 1916), 110.

13. McKinney, *Wallace and Natural Selection*, 80–96.

14. Alfred Russel Wallace, "The Origin of the Human Races and the Antiquity of Man Deduced from the Theory of Natural Selection," *Journal of Anthropological Society of London* 2 (1864): clviii–clxx.

15. Alfred Russel Wallace, "Sir Charles Lyell on Geological Climates and the Origin of Species," *London Quarterly Review* 126 (1869): 187–205.

16. Darwin, March 27, 1869, quoted in Marchant, *Alfred Russel Wallace*, 197.

17. Darwin, *The Origin*, 227–229.

18. Wallace, "Sir Charles Lyell," 205.

19. Malcom Jay Kottler, "Alfred Russel Wallace, the Origin of Man, and Spiritualism," *Isis* 65 (1974): 152.

20. Alfred Russel Wallace, "The Limits of Natural Selection as Applied to Man," *Contributions to the Theory of Natural Selection* (London: Macmillan, 1875), 356.

21. Ibid., 359-360.

22. Wallace, "The Origin of the Human Races," clxix-clxx.

23. See Charles Smith, "Alfred Russel Wallace on Spiritualism, Man and Evolution: An Analytical Essay," 1999, available at www.wku.edu/ "smithch/indexl.htm.

24. Alfred Russel Wallace, *Miracles and Modern Spiritualism* (1874; repr., London: Spiritualist Press, 1878; repr., 1955), 8.

25. Wallace, *My Life*, 2:276.

26. Wallace, *Miracles and Modern Spiritualism*, 125.

27. Ibid., 126.

28. Ibid., 127–129.

29. Ibid., 132.

30. Alfred Russel Wallace, "A Postscript to 'A New Medium,'" *Spiritual Magazine*, 2 (1867): 52, quoted in Kottler, "Alfred Russel Wallace," 169.

31. Wallace, November 12, 1866, quoted in Marchant, *Alfred Russel Wallace*, 418.

32. Thomas Huxley, November ? 1866, quoted in Marchant, *Alfred Russel Wallace*, 418.

33. Thomas Huxley, January 27, 1874, Charles Darwin, January 29, 1874, in Thomas Huxley, *Life and Letters of Thomas Huxley*, 2 vols., ed. Leonard Huxley (New York: Appleton, 1900), 1:452–456. Hereafter cited as LLTHH.

34. John Tyndall to Wallace, February 8, 1867, British Library, Add.Mss. 46439.

35. Thomas Huxley, "On the Zoological Relations of Man with the Lower Animals," *The Natural History Review* (1861): 67–84. Also in Huxley, *Scientific Memoirs of Thomas Henry Huxley*, 4 vols., ed. M. Foster and E. Ray Lancaster (London: Macmillan, 1898–1902), 2:473. Hereafter cited as SMTHH.

36. Thomas Huxley, *Man's Place in Nature* (1863; repr., New York: Appleton, 1898), 153.

37. Ibid., 152-154.

38. Thomas Huxley, *Evolution and Ethics and Other Essays* (1893; repr., London: Macmillan, 1894), 79–80.

39. I have argued elsewhere that Huxley's seemingly damning critique of evolutionary ethics must be viewed in the context of the time. He was responding to the harsh social program as advocated by Herbert Spencer that became known as social Darwinism. Examining Huxley's entire corpus of work demonstrates that his view of nature was not as harsh at it appears in the Romanes Lectures on Evolution and Ethics. See Sherrie Lyons, "Introduction," in Thomas Huxley, *Evolution and Ethics* (repr., New York: Barnes and Noble, 2007).

40. Darwin, *The Descent of Man.*

41. Ibid., 92.

42. Ibid., 67.

43. Charles Darwin, "December 19, 1832." *The Voyage of the Beagle* (1839; repr., New York: The Modern Library, 2001), 194.

44. Marchant, *Alfred Russel Wallace*, 389.

45. Ibid., Wallace to Darwin, April 18, 1869, 200.

46. Wallace, *My Life*, 2:298.

47. Thomas Huxley, "The Physical Basis of Life" (1868), *Collected Essays: Method and Results* (New York: D. Appleton, 1897), 133–134.

48. See Lightman, *The Origins of Agnosticism.*

49. Wallace, *My Life*, 2:295.

50. Alfred Russel Wallace, *The World of Life: A Manifestation of Creative Power, Directive Mind and Ultimate Purpose* (New York: Moffat, Yard, 1910. See also Martin Fichman, *An Elusive Victorian* (Chicago: University of Chicago Press, 2004).

51. Wallace, *The World of Life*, 424–425.

6. *Thatige Skepsis*

1. See Desmond, *The Politics of Evolution.*

2. Thomas Huxley, "Vestiges of the Natural History of Creation Tenth Edition London, 1853," *The British and Foreign Medico-Chirurgical Review* 13 (1854): 425-439; also see *SMTHH*, suppl., 1–19.

3. Herbert Spencer, *An Autobiography*, 2 vols. (New York: Appleton, 1904), 1:201.

4. Thomas Huxley, "On the *Origin of Species*," in LLCD 1:543-544.

5. Michael Bartholomew, "Lyell and Evolution: An Account of Lyell's Response to the Prospect of an Evolutionary Ancestry for Man," *British Journal of the History of Science* 6 (1973): 284.

6. Darwin, *The Origin*, 129, 133.

7. K. E. von Baer, "Philosophical Fragments" *Uber Entwickelungsgeschichte The Fifth Scholium*, 1828, in A Henfrey and T. Huxley, ed., *Scientific Memoirs: Natural History* (London: Taylor and Francis, 1853), 210–214.

8. Richard Owen, "Report on the Archetype and Homologies of the Vertebrate Skeleton," *BAAS Report* (1846): 175. Owen originally defined these terms in 1843 in his *Lectures on Invertebrate Animals.*

9. Darwin to Asa Gray, September 10, 1860, in LLCD, 2: 131.

10. Huxley, "On the Reception," 1:551.

11. See Mary Winsor, "The Impact of Darwinism upon the Linnaean Enterprise, with Special Reference to the Work of T. H. Huxley," in *Contemporary Perspectives on Linnaeus*, ed. John Weinstock (Lanham, MD: University Press of America, 1985), 60.

12. Huxley, "November 23, 1859," in LLTHH, 1:189.

13. Darwin, *The Origin,* 292.

14. Thomas Huxley, "The Demonstrative Evidence of Evolution," *Popular Science Monthly* 10 (1877): 294.

15. Thomas Huxley, "On the Recent Work of the 'Challenger' Expedition, and Its Bearing on Geological Problems," Notes from *Proceedings Royal Institute Great Britain* 7 (1875): 354–357; see also SMTHH, 4:64.

16. Thomas Huxley, "On the Coming of Age of The *Origin of Species*" (1880), *Collected Essays of Thomas Huxley: Darwiniana* (London: Macmillan, 1893), 242.

17. Thomas Huxley, "The Darwinian Hypothesis" (1859), *Collected Essays: Darwiniana*, 20.

18. Darwin, *The Origin,* 288, 264.

19. *Primula* is a genus of four hundred to five hundred species of low-growing herbs in the family Primulaceae that includes the primrose, cowslip and oxlip. The primrose in particular was a favorite of horticulturalists who had developed many different varieties with a wide range of different shaped and colored flowers.

20. Michael Ruse, *The Darwinian Revolution* (Chicago: University of Chicago Press, 1979; repr., 1981), 235–236.

21. See Sherrie Lyons, "Thomas Huxley: Fossils, Persistence, and the Argument from Design," *Journal of the History of Biology* 26 (1993): 545-569; and *Thomas Henry Huxley: The Evolution of a Scientist* (Amherst, NY: Prometheus Books, 1999).

22. Huxley, "On the Reception," 1:542.

23. Huxley quoting the *Vestiges*, in "Vestiges of the Natural History of Creation," 1.

24. Ibid., 5.

25. Ibid., 2.

26. For a discussion of the reception of *Vestiges*, see Rudwick, *The Meaning of Fossils*, 205–207.

27. Many scientists are guilty of this. The reasons why this does not seriously hinder science from moving toward an increasingly accurate view of the natural world will be explored in greater depth in the concluding chapter.

28. Huxley, "On Our Knowledge of the Causes of the Phenomena of Organic Nature" (1863) and "On the Coming of Age of The *Origin of Species*" (1880), *Collected Essays: Darwiniana*, 475, 228.

29. Gould and Eldredge began their paper on punctuated equilibria with Huxley's caution to Darwin: "You have loaded yourself with an unnecessary difficulty in adopting *Natura non facit saltum* so unreservedly," November 23, 1859, in LLTHH, 1:189. See Stephen Gould and Niles Eldredge, "Punctuated Equilibria: The Tempo and Model of Evolution Reconsidered," *Paleobiology* 3 (1977): 115–121.

30. See David Wake, "Comparative Terminology," *Science* 265 (1994): 268–269; S. Gilbert, J. Opitz, and R. A. Raff, "Resynthesizing Evolutionary and Developmental Biology," *Developmental Biology* 173 (1996): 363–364; and Brian Hall, *Homology: The Hierarchial Basis of Comparative Biology* (New York: Academic Press, 1984).

31. See D. W. McShea, "Complexity and Homoplasy," in *Homoplasy: The Recurrence of Similarity in Evolution*, ed., M. J. Sanderson and L. Hufford (San Diego, CA: Academic Press, 1996), 207–225; and "Possible Largest-Scale Trends in Organismal Evolution: Eight Live Hypotheses," *Annual Review of Ecology and Systematics* 29 (1998): 293–318.

7. Negotiating the Boundaries of Science

1. Bauer, *Scientific Literacy and the Myth of the Scientific Method*.

2. Thomas Gieryn, *Cultural Boundaries of Science* (Chicago: University of Chicago Press, 1999).

3. Thomas Huxley, quoted in G. Basalla, W. Coleman, and R. Kargon, eds., *Victorian Science* (New York: Anchor Books, 1970), 9.

4. W. J. T. Mitchell, *The Last Dinosaur Book* (Chicago: University of Chicago Press, 1998), 91.

5. Claudine Cohen, *The Fate of the Mammoth*, trans. William Rodamor (Chicago: University of Chicago Press, 2002), xxviii.

6. Adrienne Mayor, *The First Fossil Hunters: Paleontology in Greek and Roman Times* (Princeton, NJ: Princeton University Press, 2000).

7. Ibid., 10.

8. Ibid., 157–163.

9. The terms "myth," "legend," "folklore," and "popular belief" are often used interchangeably. There is some overlap. But according to Mayor myths usually deal with gods or divine beings to explain the origins of the natural world in a primeval past. Legends are folklore narratives that typically feature animals or human beings engaged in unusual exploits, often set in historical times and places. They may be embellished descriptions of actual events or natural facts. See *The First Fossil Hunters*, 286, n. 2.

10. Ibid., 31.

11. Paul Sereno, personal communication, December 2004.

12. Mark Norell, personal communication, October 2007.

13. Norman McLeod, personal communication, March 2005.

14. Paul Sereno, personal communication. See also Rudwick, *The Meaning of Fossils.*

15. Jack Horner, personal communication, March 2005.

16. Mark Norell, personal communication, October 2007

17. Ibid.

18. Taphonomy is the study of a decaying organism over time.

19. See Mark Norell, *Unearthing the Dragon* (New York: Pi Press, 2005); C. Brochu and M. A. Norell, "Temporal Congruence and the Origin of Birds," *Journal of Vertebrate Paleontology* 20:1 (2000): 197–200; and M. A. Norell and J. M. Clark, "Birds Are Dinosaurs," *Science Spectrum* 8 (1997): 28–34.

20. George Gaylord Simpson, "The Beginnings of Vertebrate Paleontology in North America," *Journal of Paleontology* 17:1 (1942): 26–38, quoted in Adrienne Mayor, *Fossil Legends of the First Americans* (Princeton, NJ: Princeton University Press. 2005), xxii.

21. Tim Tokaryk, personal communication, May 2005.

22. Rudwick, *Scenes from Deep Time*, 141–144. See also Adrian Desmond, "Designing the Dinosaur," *Isis* 70 (1979): 224–234.

23. Michael Freeman, *Victorians and the Prehistoric* (New Haven, CT: Yale University Press. 2004), 161.

24. Ibid., 228.

25. Rupke, *Richard Owen.*

26. Dinosaurs are not terrible in the modern sense of the word as meaning bad, but Owen used the word that at the time meant fearfully great.

27. Norell, personal communication, October 2007.

28. Sereno, quoted in Mitchell, *The Last Dinosaur Book*, 142.

29. American Museum of Natural History paleontologist Michael Novacek describing the discovery of Protoceratops fossils on a 1993 expedition to the Gobi Desert with fellow paleontologist Mark Norell. See http://www.amnh.org/exhibitiosn/mythiccreatures/land/griffin.php.

30. Norell, personal communication, October 23, 2007.

31. See Donna Haraway, *Primate Visions* (New York: Routledge, 1989).

32. For example, the notorious Piltown man, found in Sussex, England, in 1912 and was regarded as one of our earliest ancestors, was not exposed as a hoax until 1953 when Kenneth Oakley demonstrated unequivocally that the remains were a human cranium and an orangutan jaw, both of which had been artificially stained. The hoax was perpetuated for as long as it was because it fit into the researchers preconceived biases, which were that our ancestor would be big brained and not from Africa, which in turn was a reflection of their race, and country of origin. See Stephen Gould, "Piltown Revisited," *The Panda's Thumb* (New York: Norton, 1982), 108–124. See also Haraway, *Primate Visions*. For a good concise treatment of the various theories concerning human evolution, see Matt Cartmill, David Pilbeam, and Glynn Issac, "One Hundred Years of Paleoanthropology," *American Scientist* 74 (1986): 410–420.

33. L. Cosmides and J. Tooby, "Origins of Domain Specificity: The Evolution of Functional Organization," in *Mapping the Mind,* ed. L. Hirschfeld and S. Gelman (Cambridge: Cambridge University Press, 1994).

34. Stephen Gould and Richard Lewontin, "The Spandrels of San Marco and the Panglossian Paradigm: A Critique of the Adaptationist Programme," *Proceedings of the Royal Society of London* 205 (1978): 581–598.

35. See Robert Wright, *The Moral Animal* (New York: Pantheon, 1994); and Mark Ridley, *The Red Queen* (New York: Macmillan, 1993).

36. David Buss, "Sex Differences in Human Mate Preferences: Evolutionary Hypotheses Tested in 37 Cultures," *Behavioral and Brain Sciences* 12 (1989): 1-49.

37. David Buss, "Psychological Sex Differences," *American Psychologist*, March 1995, 164–168.

38. Margo Wilson and Martin Daly, "The Man Who Mistook His Wife for a Chattel," in *The Adapted Mind: Evolutionary Psychology and the Generation of Culture*, ed. J. Barkow, L. Cosmides, and J. Tooby (New York: Oxford University Press, 1992).

39. Natalie Angier, *Woman, an Intimate Geography* (Boston: Houghton Mifflin, 1999), 330–331, but see entire chapter.

40. Sarah Blaffer Hrdy, quoted in Angier, *Woman*, 331-332.

41. Sarah Blaffer Hrdy, *The Woman Who Never Evolved* (Cambridge, MA: Harvard University Press, 1981); and Jeanne Altmann, "Mate Choice and Intrasexual Reproductive Competition: Contributions to Reproduction that go Beyond Acquiring Mates," in *Feminism and Evolutionary Biology: Boundaries, Intersections, and Frontiers*, ed. Patricia A. Gowaty (New York: Chapman and Hall, 1997), 320–333.

42. Sarah Blaffer Hrdy, "Empathy Polyandry, and the Myth of the Coy Female," in *Feminist Approaches to Science*, ed. Ruth Bleier (New York: Pergamon, 1986), 119–146.

43. Steven Pinker, "Boys Will Be Boys," *The New Yorker*, February 9, 1998.

44. Jerry Fodor, *The Modularity of Mind* (Cambridge, MA: MIT Press, 1983).

45. The other three are avoidance of harm, reward dependence, and persistence.

46. R. Ebstein et al., "Dopamine D4 Receptor (D4DR) Exon III Polymorphism Associated with the Human Personality Trait of Novelty Seeking," *Nature Genetics* 12 (1996): 78–80; and J. Benjamin et al., "Population and Familial Association between the D4 Dopamine Receptor Gene and Measures of Novelty Seeking," *Nature Genetics* 12 (1996): 81–84.

47. C. R. Cloninger, D. M. Syrakic, and T. R. Przybeck, "A Psychobiological Model of Temperament and Character," *Archives of General Psychiatry*. 50 (1993): 975–990.

48. Natalie Angier, Variant Gene Tied to a Love of New Thrills," *New York Times*, January 2, 1996.

49. A. K. Malhotra et al., "The Association between the Dopamine D4 Receptor (D4Dr) 16 Amino Acid Repeat Polymorphism and Novelty Seeking," *Molecular Psychiatry* 1:5 (1996): 388–391.

50. While people do fill out personality questionnaires differently, these differences do not correlate well to behavioral differences. Rather, the vast majority of variance can be explained by (1) situational differences, (2) the persons's construal of the situation, and (3) the person's goals. See Richard Nisbett and Lee Ross, *Human Inference: Strategies and Shortcomings of Social Judgment* (Englewood Cliffs, NJ: Prentice Hall, 1980); and Lee Ross and Richard Nisbett, *The Person and the Situation* (Philadelphia: Temple University Press, 1991).

51. Jerome Kagan, *Galens Prophecy* (New York: Basic Books, 1994).

52. Robert Plomin, *Genetics and Experience* (Thousand Oaks, CA: Sage, 1994).

53. Todd Granthan and Shawn Nichols, "Evolutionary Psychology: Ultimate Explanations and Panglossian Predictions," in *Where Biology Meets Psychology: Philosophical Essays*, ed. Valorie Hardcastle (Cambridge, MA: MIT Press, 1999), 47–66.

54. Gould, *The Mismeasure of Man.*

55. Evelyn Fox-Keller, "Nature, Nurture, and the Human Genome Project," in *The Code of Codes,* ed. Daniel Kevles and Leroy Hood (Cambridge, MA: Harvard University Press, 1992).

56. John McCarter, CEO and president of the Field Museum in Chicago maintains that good exhibits use artifacts to tell a story that

promotes a particular point of view. See Kathy A. Svitil, "The Discover Interview," *Discover Magazine*, May 2006, 54–57.

57. Nucleic acid sequence data has led to the classification of all life-forms into three domains: eukaryotes, bacteria and Archaea, although not everyone accepts this view. For a history of this shift and also the controversy surrounding the classification of Archaea as a third domain, see Sherrie Lyons, "Thomas Kuhn Is Alive and Well: The Evolutionary Relationships of Simple Life Forms, A Paradigm Under Siege?" *Perspectives in Biology and Medicine* 45:3 (Summer 2002): 359–376.

58. Allan Wilson, quoted by Stephen Gould in Cohen, *The Fate of the Mammoth*, xvii.

59. Douglas Allchin, "Error Types," *Perspectives on Science* 9 (2001): 38–59.

Bibliography

Agassiz, Louis. "1849 Lecture," in *A Romance of the Sea Serpent or the Ichthyosaurus and a Collection of Ancient and Modern Authorities*, ed. Eugene Bachelder (Cambridge: John Bartlett, 1849), 135–136.

Allchin, Douglas. "Error Types." *Perspectives on Science* 9 (2001): 38–59.

Altmann, Jeanne. "Mate Choice and Intrasexual Reproductive Competition: Contributions to Reproduction that go Beyond Acquiring Mates." In *Feminism and Evolutionary Biology: Boundaries, Intersections, and Frontiers*, edited by Patricia A. Gowaty. New York: Chapman and Hall, 1997.

Angier, Natalie. "Variant Gene Tied to a Love of New Thrills." *New York Times*, January 2, 1996.

———. *Woman: An Intimate Geography.* Boston, MA: Houghton Mifflin, 1999.

Baer, Karl Ernst von. "Philosophical Fragments *Uber Entwickelungsgeschichte The Fifth Scholium.*" 1828. In *Scientific Memoirs: Natural History*, edited by A. Henfrey and T. Huxley. London: Taylor and Francis, 1853.

Barrett, Paul, ed. *Metaphysics, Materials and the Evolution of Mind.* Chicago: University of Chicago Press, 1974.

Barthlomew, Michael. "Huxley's Defense of Darwin." *Annals Science* 32 (1975): 525–535.

———. "Lyell and Evolution: An Account of Lyell's Response to the Prospect of an Evolutionary Ancestry for Man." *British Journal of the History of Science* 6 (1973): 261–303.

———. "The Non-Progress of Non Progression: Two Responses to Lyell's Doctrine." *British Journal of the History of Science* (1976): 166–174.

———. "The Singularity of Lyell." *History of Science* 17 (1979): 276–293.

Basalla, G., W. Coleman, and R. Kargon, eds. *Victorian Science.* New York: Anchor Books, 1970.

Batchelder, Eugene, ed. *A Romance of the Sea Serpent or the Ichthyosaurus and a Collection of Ancient and Modern Authorities.* Cambridge: John Bartlett, 1849.

Bauer, Henry. *Scientific Literacy and the Myth of the Scientific Method.* Chicago: University Illinois Press, 1992.

Beloff, John. "Home's Levitations." *Skeptical Inquirer* 10 (1986): 370–371.

Benjamin, J. et al. "Population and Familial Association between the D4 Dopamine Receptor Gene and Measures of Novelty Seeking." *Nature Genetics* 12 (1996): 81–84.

Black, Barbara J. *On Exhibit, Victorians and Their Museums.* Charlottesville: University Press of Virginia, 2000.

Blaffer Hrdy, Sarah. "Empathy Polyandry, and the Myth of the Coy Female." In *Feminist Approaches to Science*, edited by Ruth Bleier. New York: Pergamon, 1986.

———. *Mother Nature.* New York: Pantheon Books, 1999.

———. *The Woman Who Never Evolved.* Cambridge, MA: Harvard University Press, 1981.

Bowler, Peter. "Darwinism and the Argument from Design." *Journal of the History of Biology* 10 (1977): 29–43.

———. *Evolution.* Berkeley and Los Angeles: University of California Press, 1984.

———. *Fossils and Progress.* New York: Science History Publications, 1976.

Brochu, C., and M. A. Norell. "Temporal Congruence and the Origin of Birds." *Journal of Vertebrate Paleontology* 20:1 (2000): 197-200.

Brooke, John H. *Science and Religion: Some Historical Perspectives.* Cambridge: Cambridge University Press, 1991.

Browne, Janet. *Charles Darwin: Voyaging.* Princeton, NJ: Princeton University Press, 1995.

Buckland, William. *Geology and Mineralogy Considered with Reference to Natural Theology.* Philadelphia: Phillip Carey, Lea and Blanchard, 1837.

———. *Reliquiae Diluvianae.* London: John Murray, 1823.

Bullough, Vernon. "Spirit Rapping Unmasked: An 1851 Investigation and Its Aftermath." *The Skeptical Inquirer* 10 (1985): 61–67.

Buss, David. "Psychological Sex Differences." *American Psychologist,* March 1995, 164–168.

———. "Sex Differences in Human Mate Preferences: Evolutionary Hypotheses Tested in 37 Cultures." *Behavioral and Brain Sciences* 12 (1989): 1–49.

Cannon, Walter Cannon. "The Uniformitarian-Catastrophist Debate." *Isis* 51 (1960): 38–55.

Cantor, Geoffrey N. "The Edinburgh Phrenology Debate: 1802–1828." *Annals of Science* 32 (1975): 195–218.

Cartmill, Matt, David Pilbeam, and Glynn Issac. "One Hundred Years of Paleoanthropology." *American Scientist* 74 (1986): 410–420.

(Chambers, Robert). *Vestiges of the Natural History of Creation.* 1844. Reprint, Surrey, UK: Leicester University Press, 1969.

"Charles Darwin, the Eminent Naturalist." *American Phrenological Journal and Life Illustrated* 48:4 (October 1868): 121–122.

Clarke, E., and L. S. Jacyna. *Nineteenth-Century Origins of Neuroscientific Concepts.* Berkeley and Los Angeles: University of California Press, 1987.

Cloninger, C. R., D. M. Syrakic, and T. R. Przybeck. "A Psychobiological Model of Temperament and Character." *Archives of General Psychiatry* 50 (1993): 975–990.

Cogswell, C. "The Great Sea-Serpent." *Zoologist* 6 (1848): 2316–2323.

Cohen, Claudine. *The Fate of the Mammoth*, translated by William Rodamor. Chicago: University of Chicago Press, 2002.

Coleman, William. *Biology in the Nineteenth Century: Problems of Form, Function, and Transmutation.* Cambridge: Cambridge University Press, 1977.

———. *George Cuvier, Zoologist.* Cambridge, MA: Harvard University Press, 1964.

———. "Lyell and the Reality of Species." *Isis* 53 (1962): 325–338.

———. "Morphology between Type Concept and Descent Theory." *Journal of History of Medicine* 31 (1976): 149–175.

Combe, George. *The Constitution of Man.* 1828. Reprint, Boston, MA: Marsh, Capen, and Lyon, 5th ed., 1835.

Conybeare, William. "On the Discovery of an Almost Perfect Skeleton of Plesiosaurus." Transactions of the Geological Society of London, 2nd series, 1824.

Cosmides, L., and J. Tooby. "Origins of Domain Specificity: The Evolution of Functional Organization." In *Mapping the Mind*, edited by L. Hirschfeld and S. Gelman. Cambridge: Cambridge University Press, 1994.

Cooter, Roger. *The Cultural Meaning of Popular Science.* New York: Cambridge University Press, 1984.

———. *Phrenology in the British Isles.* Metuchen, NJ: Scarecrow Press, 1989.

Crafts, William. *The Sea Serpent or Gloucester Hoax: A Dramatic Jeu d'Espirit in Three Acts.* Charleston, SC: A. E. Miller, 1819.

Critchley, MacDonald. *The Divine Banquet of the Brain and Other Essays.* New York: Raven Press, 1979.

Crookes, William. "Spiritualism Viewed by the Light of Modern Science." *Quarterly Journal of Science,* July 1870.

Damasio, Antonio. *Descartes' Error.* New York: Avon Books, 1994.

Darlington, C. D. *The Facts of Life*. London: Allen and Unwin, 1953.

Darwin, Charles. *Animals and Plants Under Domestication*, 2nd ed. New York: Appleton, 1892.

———. *The Autobiography of Charles Darwin and Selected Letters*, edited by Francis Darwin. New York: Dover, 1958.

———. *The Descent of Man and Selection in Relation to Sex*. London: John Murray, 1871. Princeton, NJ: Princeton University Press, 1981.

———. *The Expression of the Emotions in Man and Animals*. 1872; repr., Chicago: University of Chicago Press, 1965.

———. *Life and Letters of Charles Darwin*, 2 vols, edited by Francis Darwin. London: John Murray, 1887.

———. *More Letters of Charles Darwin*, 2 vols, edited by Francis Darwin and A. C. Seward. London: John Murray, 1903.

———. *On The Origin of Species by Means of Natural Selection*. London: John Murray, 1859; New York: Avenel Books, 1976.

———. *The Voyage of the Beagle*, 1839. New York: The Modern Library, 2001.

Desmond, Adrian. *Archetypes and Ancestors Paleontology in Victorian London 1850–1875*. Chicago: University of Chicago Press, 1982; repr., 1984.

———. "Designing the Dinosaur." *Isis* 70 (1979): 224–234.

———. *The Politics of Evolution: Morphology, Medicine, and Reform in Radical London*. Chicago: University of Chicago Press, 1989.

Eberhart, G. M. *Monsters*. NewYork: Garland, 1983.

Ebstein, R. et al. "Dopamine D4 Receptor (D4DR) Exon III Polymorphism Associated with the Human Personality Trait of Novelty Seeking." *Nature Genetics* 12 (1996): 78–80.

Eliot, George. *George Eliot's Life as Related in Her Letters and Journals*, 3 vols, edited by J. W. Cross. New York: Harper Brothers, 1885.

Elliotson, John. "Reply to the Attacks on Phrenology." *Lancet* (November 26, December 10, 1831): 287–294, 357–364.

Ellis, Richard. *Monsters of the Sea*. New York: Knopf, 1994.

———. *Sea Dragons, Predators of the Prehistoric Seas*. Lawrence: University Press of Kansas, new edition, 2005.

Farber, Paul. "The Type-Concept in Zoology." *Journal of the History of Biology* 9 (1976): 93–119.

Fichman, Martin. *An Elusive Victorian*. Chicago: University of Chicago Press, 2004.

Fletcher, J. *Rudiments of Physiology*, edited by Robert Lewins. Edinburgh: John Carfrad and Son, 1837.

Fodor, Jerry. *The Modularity of Mind*. Cambridge, MA: MIT Press, 1983.

Fox-Keller, Evelyn. "Nature, Nurture, and the Human Genome Project." In *The Code of Codes*, edited by Daniel Kevles and Leroy Hood. Cambridge, MA: Harvard University Press, 1992.

Freeman, Michael. *Victorians and the Prehistoric.* New Haven, CT: Yale University Press, 2004.

Frye, Northrop. *Words with Power.* New York: Harcourt Brace Jovanovich, 1990.

Gardner, Marvin. *Fads and Fallacies in the Name of Science.* New York: Putnam's, 1952.

Gieryn, Thomas. *Cultural Boundaries of Science.* Chicago: University of Chicago Press, 1999.

Gilbert, S., J. Opitz, and R. A. Raff. "Resynthesizing Evolutionary and Developmental Biology." *Developmental Biology* 173 (1996): 357–372.

Gillispie, Charles. *Genesis and Geology: A Study in the Relations of Scientific Thought, Natural Theology and Social Opinions in Great Britain, 1790–1850.* 1951. Reprint, New York: Harper, 1959.

Gordon, John, ed. *A Collection of Scientific Articles and Newspaper Reports from 1816-1905 Relating to the Sea Serpent.* South Brewer, ME: John Gordon, 1926.

Gosse, Phillip Henry. *Romance of Natural History.* London, 1860.

Gould, Stephen. "Agassiz's Marginalia in Lyell's *Principles* or the Perils of Uniformity and the Ambiguity of Heroes." *Studies in the History of Biology* 3 (1979): 119–138.

———. *Bully for Brontosaurus.* New York: Norton, 1991.

———. "The Eternal Metaphors of Paleontology." In *Patterns of Evolution,* edited by A. Hallam. Amsterdam: Elsevier, 1977.

———. *The Mismeasure of Man.* New York: Norton, 1982.

———. "Piltown Revisited." In *The Panda's Thumb.* New York: Norton, 1982.

Gould, Stephen, and Niles Eldredge. "Punctuated Equilibria: The Tempo and Model of Evolution Reconsidered." *Paleobiology* 3 (1977): 115–121.

Gould, Stephen, and Richard Lewontin. "The Spandrels of San Marco and the Panglossian Paradigm: a Critique of the Adaptationist Programme." *Proceedings of the Royal Society of London* 205 (1978): 581–598.

Granthan, Todd, and Shawn Nichols. "Evolutionary Psychology: Ultimate Explanations and Panglossian Predictions." In *Where Biology Meets Psychology: Philosophical Essays*, edited by Valorie Hardcastle. Cambridge, MA: MIT Press.

Hall, Brian. *Homology: The Hierarchical Basis of Comparative Biology.* New York: Academic Press, 1984.

Hall, Trevor. *The Engima of Daniel Home*. Buffalo, NY: Prometheus Books, 1984.

———. *The Spiritualists*. 1962. Reprinted as *The Medium and The Scientist*. Buffalo, NY: Prometheus Books, 1984.

Hall, Trevor, and Ian Stevenson. *Journal of the American Society for Psychial Research* 57:4 (1963): 215–226; 58:1 (1964): 57–65; 58:2 (1964): 128–133.

Hallam, A. ed. *Patterns of Evolution*. Amsterdam: Elsevier, 1977.

Hanen, M. P., M. Osler, and R. G. Weyant, eds. *Science, Pseudo-Science, and Society*. Waterloo, ON: Wilfrid Laurier University Press, 1980.

Haraway, Donna. *Primate Visions*. New York: Routledge, 1989.

Henfrey, A., and T. Huxley, eds. *Scientific Memoirs: Natural History*. London: Taylor and Francis, 1853.

Heuvelmans, Bernard. *In the Wake of Sea-Serpents*, translated by Richard Garnet. New York: Hill and Wang, 1968.

Hodkins, Timothy. "August 15, 1818," in *A Collection of Scientific Articles and Newspaper Reports from 1816-1905 Relating to the Sea Serpent*, ed. John Gordon. South Brewer, ME: John Gordon, 1926.

Home, Daniel Douglas. *Lights and Shadows of Spiritualism*. London: Virtue, 1877.

Hooykaas, R. *Natural Law and Divine Miracle: The Principle of Uniformity, Biology, and History*. Leiden, Netherlands: Brill, 1959.

Houghton, Walter. *The Victorian Frame of Mind, 1830–1870*. 1957. Reprint, New Haven, CT: Yale University Press, 1985.

Hudson, D., and B. Profet. "A Bumpy Start to Science Education." *New Scientist* (August 14, 1986): 25–28.

Hutton, James. *Theory of the Earth. Transactions Royal Society of Edinburgh* 1, 1788.

Huxley, Thomas. *Collected Essays:*

Vol. 1: *Method and Results*. New York: Appleton, 1897.

Vol. 2: *Darwiniana*. London: Macmillan, 1893.

Vol. 6: *Hume*. London: Macmillan, 1897.

Vol. 7: *Man's Place in Nature*. New York: Appleton, 1863. Reprint, 1898.

Vol. 9: *Evolution and Ethics and Other Essays*. London: Macmillan, 1894.

———. "The Demonstrative Evidence of Evolution." *Popular Science Monthly* 10 (1877): 285–298.

———. *Evolution and Ethics and Other Essays*. 1893. Introduction by Sherrie Lyons. New York: Barnes and Noble, 2006.

———. *Life and Letters of Thomas Huxley*, 2 vols, edited by Leonard Huxley. New York: Appleton, 1900.

———. "On the Animals Which Are Most Nearly Intermediate between Birds and Reptiles." *Scientific Memoirs of Thomas Henry Huxley*, 4 vols., ed. M. Foster and E. Ray Lancaster. London: Macmillan, 1898–1902, vol. 3. Hereafter SMTHH.

———. "On the Recent Work of the 'Challenger' Expedition, and Its Bearing on Geological Problems." Notes from *Proceedings Royal Institute Great Britain* 7 (1875): 354–357. Also in SMTHH 4:64.

———. "On the Zoological Relations of Man with the Lower Animals." *The Natural History Review* (1861): 67–84. Also in SMTHH 2:471–492.

———. "The Reception of the 'Origin of Species.'" In *Life and Letters of Charles Darwin,* 2 vols, edited by Francis Darwin. London: John Murray, 1887.

———. *Scientific Memoirs of Thomas Henry Huxley*, 4 vols, edited by M. Foster and E. Ray Lancaster. London: Macmillan, 1898–1902.

———. "Vestiges of the Natural History of Creation Tenth Edition London, 1853." *The British and Foreign Medico-Chirurgical Review* 13 (1854): 425–439.

Kagan, Jerome. *Galens Prophecy*. New York: Basic Books, 1994.

Knell, Simon. *The Culture of English Geology, 1815–1851*. Burlington, VT: Ashgate, 2000.

Knight, David. *The Age of Science*. New York: Blackwell, 1986.

Kottler, Malcom Jay. "Alfred Russel Wallace, the Origin of Man, and Spiritualism." *Isis* 65 (1974): 145–192.

Kuhn, Thomas. *The Structure of Scientific Revolutions*. Chicago: University of Chicago Press, 1962. Reprint, 1970.

Lauden, Rachel, ed. *The Demarcation between Science and Pseudo-Science*, vol. 2. Blacksburg: Center for the Study of Science in Society, Virginia Polytechnic Institute, 1983.

Lightman, Bernard. *The Origins of Agnosticism: Victorian Unbelief and the Limits of Knowledge*. Baltimore: Johns Hopkins University Press, 1987.

———, ed. *Victorian Science in Context*. Chicago: University of Chicago Press, 1997.

Lyell, Charles. "Anniversary Address of the President." *Quarterly Journal of the Geological Society* 7 (1851): xxv–lxxvi.

———. Charles Lyell Scientific Correspondence, Edinburgh Library, Box 26.

———. *Life and Letters of C. Lyell*, 2 vols, edited by K. Lyell. London: John Murray, 1881.

———. *Principles of Geology: Being an Inquiry How Far the Former Changes of the Earth's Surface Are Referable to Causes Now in*

Operation, 3 vols. London: John Murray, 1830–1833. Reprint, Chicago: University of Chicago Press, 1990.

———. *A Second Visit to the United States of North America,* 2 vols. London: John Murray, 1847.

Lyons, Sherrie. "At the Margins of Science: Sea Serpent Investigations in the Victorian Era." In *Repositioning Victorian Science*, edited by David Clifford, Elizabeth Wedge, Alex Warwick and Martin Willis. London: Anthem Press, 2006.

———. "The Origins of T. H. Huxley's Saltationalism: History in Darwin's Shadow." *Journal of the History of Biology* 28 (1995): 463–494.

———. "Sea Monsters: Myth or Genuine Relic of the Past?" In *Oceanographic History*, edited by K. Benson and P. Rehbock. Seattle: University of Washington Press, 2002.

———. "Science or Pseudo-Science: Phrenology as a Cautionary Tale for Evolutionary Psychology." *Perspectives in Biology and Medicine* 41:4 (Summer 1998): 491–503.

———. *Thomas Henry Huxley: The Evolution of a Scientist.* Amherst, NY: Prometheus Books, 1999.

———. "Thomas Huxley: Fossils, Persistence, and the Argument from Design." *Journal of the History of Biology* 26 (1993): 545–569.

———. "Thomas Kuhn Is Alive and Well: The Evolutionary Relationships of Simple Life Forms, A Paradigm Under Siege?" *Perspectives in Biology and Medicine* 45:3 (Summer 2002): 359–376.

Malhotra, A. K. *et al.* "The Association between the Dopamine D4 Receptor (D4Dr) 16 Amino Acid Repeat Polymorphism and Novelty Seeking." *Molecular Psychiatry* 1:5 (1996): 388–391.

Mantell, Gideon. *The Wonders of Geology,* 2 vols. London: Relfe and Fletcher, Cornhill, 1838.

Marchant, James. *Alfred Russel Wallace, Letters and Reminiscences.* New York: Harper Brothers, 1916.

Mauskopf, Seymour. "Marginal Science." In *Companion to the History of Modern Science,* edited by R. C. Olby, G. N. Cantor, J. R. R. Christie, and M. J. S. Hodge. London: Routledge, 1990.

Mayor, Adrienne. *The First Fossil Hunters: Paleontology in Greek and Roman Times*. Princeton, NJ: Princeton University Press. 2000.

———. *Fossil Legends of the First Americans*. Princeton, NJ: Princeton University Press, 2005.

McKinney, H. Lewis. "Alfred Russel Wallace and the Discovery of Natural Selection." *Journal of History of Medicine* 21 (1966): 333–357.

———. *Wallace and Natural Selection*. New Haven, CT: Yale University Press, 1972.

McShea, D. W. "Complexity and Homoplasy." In *Homoplasy: the Recurrence of Similarity in Evolution*, edited by M. J. Sanderson and L. Hufford. San Diego, CA: Academic Press, 1996.

———. "Possible Largest-Scale Trends in Organismal Evolution: Eight Live Hypotheses." *Annual Review of Ecology and Systematics* 29 (1998): 293–318.

Mitchell, W. J. T. *The Last Dinosaur Book*. Chicago: University of Chicago Press, 1998.

Morrell, Jack, and Arnold Thackray. *Gentlemen of Science*. Oxford, UK: Clarendon Press, 1981.

Newman, Edward. "The Great Sea Serpent." *Westminster Review* 50 (1849): 491–515.

Nielsen, Jorgen, and V. Larsen. "Remarks on the Identity of the Giant Dana Eel-Larva." *Vidensk. Meddr dansk naturh. Foren* 133 (1970): 149–157.

Nisbett, Robert, and Lee Ross. *Human Inference: Strategies and Shortcomings of Social Judgment*. Englewood Cliffs, NJ: Prentice Hall, 1980.

Noel, P., and E. Carlson. "Origins of the Word 'Phrenology.'" *American Journal of Psychiatry* 127 (1970): 694–697.

Norell, Mark. *Unearthing the Dragon*. New York: Pi Press, 2005.

Norell, M. A., and J. M. Clark. "Birds Are Dinosaurs." *Science Spectrum* 8 (1997): 28–34.

Oppenheim, Janet. *The Other World*. Cambridge: Cambridge University Press, 1985.

Ospovat, Dov. *The Development of Darwin's Theory: Natural History, Natural Theology, and Natural Selection, 1838–1839*. Cambridge: Cambridge University Press, 1981.

———. "The Influence of Karl Ernst von Baer's Embryology, 1828–1859: A Reappraisal in Light of Richard Owen and William B. Carpenter's Paleontological Application of von Baer's Law." *Journal of the History of Biology* 9 (1976): 1–28.

———. "Lyell's Theory of Climate." *Journal of the History of Biology* 10 (1977): 317–339.

Owen, Richard. "The Great Sea Serpent." *The Annals and Magazine of Natural History*, 2nd ser., 2 (1848): 458–463.

———. *Paleontology*. Edinburgh: Black, 1861.

———. "Report on the Archetype and Homologies of the Vertebrate Skeleton." *BAAS Report* (1846): 169–340.

Paine, Thomas. *The Age of Reason: Being an Investigation of True and Fabulous Theology*. Paris, 1794.

Paley, William. *Natural Theology*. 1802. Reprint, London: Farnbourgh, Gregg, 1970.

Palfreman, Jon. "Between Skepticism and Credulity: A Study of Victorian Scientific Attitudes to Modern Spirituality." In *On the Margins of Science: The Social Construction of Rejected Knowledge*. Sociological Review Monograph 27, edited by Roy Wallis. Keele, UK: University of Keele, 1979.

Paradis, James, and G. C. Williams, eds. *T. H. Huxley's "Evolution and Ethics": With New Essays on Its Victorian and Sociobiological Context*. Lawrenceville, NJ: Princeton University Press, 1989.

Park, Katherine, and Lorraine Daston. "Unnatural Conceptions: The Study of Monsters in Sixteenth and Seventeenth-Century France and England." *Past and Present* 92 (1981): 21–54.

Parrini, Paolo, Wesley C. Salmon, and Merrilee H. Salmon, eds. *Logical Empiricism—Historical and Contemporary Perspectives*. Pittsburgh: University of Pittsburgh Press, 2003.

Phrenological Journal and Magazine of Moral Science, The. XII 1839, XIV 1849.

Pinker, Steven. "Boys Will Be Boys." *The New Yorker*, February 9, 1998.

Plomin, R. *Genetics and Experience*. Thousand Oaks, CA: Sage, 1994.

Podmore, Frank. *Modern Spiritualism: A History and Criticism*, 2 vols. London: Methuen, 1902.

Popper, Karl. *The Logic of Scientific Discovery*. London: Routledge Classics, 1959. Reprint, 2002.

Richards, Robert. *Darwin and the Emergence of of Evolutionary Theories of Mind and Behavior*. Chicago: University of Chicago Press, 1987.

Ridley, Matt. *The Red Queen*. New York: Macmillan, 1993.

Ritvo, Harriet. *The Platypus and the Mermaid and Other Figments of the Classifying Imagination*. Cambridge, MA: Harvard University Press, 1997.

______. "Professional Scientists and Amateur Mermaids: Beating the Bounds in Nineteenth-Century Britain." *Victorian Literature and Culture* 19 (1991): 277–289.

Ross, Lee, and Robert Nisbett. *The Person and the Situation*. Philadelphia: Temple University Press, 1991.

Rudwick, Martin. *The Meaning of Fossils*. New York: Neale Watson Academic Publishers, 1972. Chicago: University of Chicago Press, 1982.

———. *Scenes from Deep Time: Early Pictorial Representations of the Prehistoric World*. Chicago: University of Chicago Press, 1992.

———. "Uniformity and Progression: Reflections on the Structure of Geological Theory in the Age of Lyell." In *Perspectives in the History of Science and Technology*, edited by Duane H. Roller. Norman: University of Oklahoma Press, 1971.

Rupke, Nicolaas. *Richard Owen: Victorian Naturalist.* New Haven, CT: Yale University Press, 1994.

Ruse, Michael. *The Darwinian Revolution.* Chicago: University of Chicago Press, 1979. Reprint, 1981.

Secord, James, ed. Introduction. In *Vestiges of the History of Creation and other Evolutionary Writings* (Robert Chambers). Chicago: University of Chicago Press, 1994.

———. *Victorian Sensation.* Chicago: University of Chicago Press, 2000.

Shapin, Steven. "Phrenological Knowledge and the Social Structure of Early Nineteenth-Century Edinburgh." *Annals of Science* 32 (1975): 195–218.

———. "The Politics of Observation: Cerebral Anatomy and Social Interests in the Edinburgh Phrenology Disputes." In *On the Margins of Science: The Social Construction of Rejected Knowledge,* edited by R. Wallis. Keele, UK: University of Keele, 1979.

Smith, Charles. "Alfred Russel Wallace on Spiritualism, Man and Evolution: An Analytical Essay." 1999. Available at www.wku.edu/"smithch/indexl.htm.

Spencer, Herbert. *An Autobiography,* 2 vols. New York: Appleton, 1904.

———. *Principles of Psychology: A System of Synthetic Philosophy,* 2nd ed., 2 vols. London: Williams and Norgate, 1872.

———. *Social Statics: Or the Conditions Essential to Human Happiness Specified and the First of Them Developed.* London: Chapman, 1851.

Stein, Gordon. *The Sorcerer of Kings.* Amherst, NY: Prometheus Books, 1993.

Svitil, Kathy A. "The Discover Interview." *Discover Magazine,* May 2006.

Temkin, Owsei. "Gall and the Phrenological Movement." *Bulletin of the History of Medicine* 21 (1947): 275–321.

Turner, Frank. *Between Science and Religion.* New Haven: Yale University Press, 1974.

Verne, Jules. *The Complete Twenty Thousand Leagues Under the Sea,* translation, introduction and notes by Emanuel J. Mickel. 1870. Reprint, Bloomington: Indiana University Press, 1969.

Wake, David. "Comparative Terminology." *Science* 265 (1994): 268–269.

Wallace, Alfred Russel. *Darwinisim: An Exposition of the Theory of Natural Selection with Some of Its Applications.* London: Macmillan, 1889.

———. "The Limits of Natural Selection as Applied to Man." *Contributions to the Theory of Natural Selection.* London: Macmillan, 1875.

———. *The Malay Archipelago.* 1869. Reprint, New York: Dover, 1962.

———. *Miracles and Modern Spiritualism.* London: Spiritualist Press, 1878. Reprint, 1955.

———. *My Life: A Record of Events and Opinions*, 2 vols. London: Chapman and Hall, 1905.

———. "On the Law Which Has Regulated the Introduction of New Species." *Annals and Magazine Natural History*, 2d ser., 16 (1855): 184–196.

———. "On the Tendency of Varieties to Depart Indefinitely from the Original Type." *Journal of the Proceedings of the Linnaean Society, Zoology* 3 (August 20, 1858): 53–62. Reprinted in Alfred Wallace, *Natural Selection and Tropical Nature: Essays on Descriptive and Theoretical Biology*. Westmead, UK: Gregg International, 1969, 20–30.

———. "The Origin of the Human Races and the Antiquity of Man Deduced From the Theory of Natural Selection." *Journal of the Anthropological Society of London* 2 (1864): clvii–clxxxvii.

———. "Sir Charles Lyell on Geological Climates and the Origin of Species." *Quarterly Review* 126 (1869): 187–205.

———. *The Wonderful Century*. London: Swan Sonnenschein, 1898.

———. *The World of Life: A Manifestation of Creative Power, Directive Mind and Ultimate Purpose*. New York: Moffat, Yard, 1910.

Westrum, Ron. "Knowledge about Sea Serpents." In *On the Margins of Science: The Social Construction of Rejected Knowledge*. Sociological Review Monograph 27, edited by Roy Wallis. Keele, UK: University of Keele, 1979.

Wilson, Leonard. *Charles Lyell: the years to 1841*. New Haven, CT: Yale University Press, 1972.

———, ed. *Sir Charles Lyell's Scientific Journals on the Species Question*. New Haven, CT: Yale University Press, 1970.

Wilson, Margo, and Martin Daly. "The Man Who Mistook His Wife for a Chattel." In *The Adapted Mind: Evolutionary Psychology and the Generation of Culture*, edited by J. Barkow, L. Cosmides, and J. Tooby. New York: Oxford University Press, 1992.

Winsor, M. "The Impact of Darwinism upon the Linnaean Enterprise, with Special Reference to the Work of T. H. Huxley." In *Contemporary Perspectives on Linnaeus*, edited by John Weinstock. Lanham, MD: University Press of America, 1985.

Winter, Alison. *Mesmerized*. Chicago: University of Chicago Press, 1998.

Wright, Robert. *The Moral Animal*. New York: Pantheon, 1994.

Yanni, Carla. *Nature's Museums*. Baltimore: John Hopkins University Press, 1999.

Young, Robert. *Mind, Brain and Adaptation in the Nineteenth Century*. Oxford, UK: Clarendon Press, 1970.

Index